Economic Growth and the Nanotechnology Priority

Economic Growth and the Nanotechnology Priority

The Challenge of Economic Crisis and Industrial Concentration

James A. Harris Sr.

LEXINGTON BOOKS
Lanham • Boulder • New York • London

Published by Lexington Books
An imprint of The Rowman & Littlefield Publishing Group, Inc.
4501 Forbes Boulevard, Suite 200, Lanham, Maryland 20706
www.rowman.com

16 Carlisle Street, London W1D 3BT, United Kingdom

British Library Cataloguing in Publication Information Available

Library of Congress Cataloging-in-Publication Data Available

ISBN 978-0-7391-9998-5 (cloth : alk. paper) ISBN 978-0-7391-9999-2 (electronic)

∞™ The paper used in this publication meets the minimum requirements of American National Standard for Information Sciences—Permanence of Paper for Printed Library Materials, ANSI/NISO Z39.48-1992.

Printed in the United States of America

To my wife, Ursula Harris, my son, James A. Harris Jr., and my son, Edmond F. Harris (deceased), with all my heart, Great Thanks for the love and eternal inspiration, encouragement, and Strength that I have received from you to make this book a contribution to the improvement of living conditions for people. To my mother, Mrs. Annie Eula Harris, and my father, Mr. Ambrose E. Harris Sr. (deceased), who ordered me from my birth "to be somebody and help other people." I give my everlasting thank-you.

Contents

1

Essence of Economic Growth–Based Nanotechnology Innovation Worldwide

The cornerstone of this book is the assessment by the author of the towering necessity for economic growth based, determined, and driven nanotechnology innovation worldwide, among which are materials nanoarchitectonics, nanosystems, nanofabrication, nanodevices, nanobio-field, nanomaterials, and nanoscale integration, and the application to the transformation and convergence of the total industries in the targeted North American Industry Classification System (NAICS) United States 2012—the twenty Sectors, into (1) the creation and construction of a national nanotechnology readiness system of government-sponsored enterprises due to enabling revolutionary nanotechnology innovation integration; (2) the establishment of global competitive advantage-driven nanotechnologies originating from the NAICS; and (3) nanotech U.S. contribution to the achievement of substantial, persistent, and sustained increases in the double-digit growth of the U.S. economy, industrial diversification expansion, and increase in demand for employment.

Within the author's assessment there is his principal desideratum and his distinguished characterization that manifest itself in major U.S. corporations' market structure, conduct, and performance; large banks; the Gramm-Leach-Bliley Act (1999) regarding Financial Bank Holding Companies; Federal Reserve Bank Certified Bank Holding Companies and their concentration of control and interlocking directorates performance, in effect, that have influenced the overall direction in the U.S. economy and accounted for market dominance in the global over-the-counter (OTC) derivatives' gigantic accumulation, huge volume of trading, and U.S. money stock measures' devastating decrease-in-demand crises, while taking action in concert and engaged in unreasonable restraint of trade. The crises have been documented in two earlier publications: (1) *The Financial Crisis Inquiry Report, Final Report of the*

National Commission on the Causes of the Financial and Economic Crisis in the United States submitted by the Financial Crisis Inquiry Commission pursuant to Public Law 111-21, January 2011, Official Government Edition; and (2) *Wall Street and the Financial Crisis: Anatomy of a Financial Collapse, Report and Appendix before the Permanent Subcommittee on Investigations of the Committee on Homeland Security and Governmental Affairs*, U.S Senate, 112th Congress, 1st Session, vol. 5, part III, April 13, 2011, Washington, DC: U.S. Government Printing Office 2011. Real gross domestic product (GDP), defined as the output of goods and services produced by labor and property located in the United States, decreased at an annual rate of 0.1 percent from the third quarter to the fourth quarter, according to the "advanced" estimate released by the Bureau of Economic Analysis (www.bea.gov). When examined in a historical perspective the slow growth of the U.S. gross domestic product (GDP), wherein current-dollar GDP is the market value of the nation's output of goods and services and real GDP is the output of goods and services produced by labor and property located in the United States, did in 2011 amount to $15,087.7 billion and in year 2000 leveled off at $9,951.5 billion.[1] This change in the level of GDP is representative of an increase of only $5,136.2 billion, an increase in the GDP that covers the 2000 to 2011 period; in effect, this pattern can be explained by first, the huge magnitude of obsolescence, deterioration, and old-fashioned technology which are embedded in NAICS industry classifications such as Manufacturing Sector 31-33 and Utilities Sector 22. The Manufacturing Sector 31-33 activities are the mechanical, physical, or chemical transformation of material, substances, or components into new products. The Utilities Sector 22 activities are generating, transmitting and/or distributing electricity, gas, steam, and water, and removing storage through a permanent infrastructure of lines, mains, and pipes. Second, the underperforming gross private domestic investment component of GDP ($1,866.4 billion in 2011 and $1,712.2 billion in 2000) and government consumption expenditures and gross investment component of GDP ($3,029.7 billion in 2011 and $1,731.0 billion in 2000). Finally, the noncompetitive exports or net exports of goods and services ($−578.2 billion in 2011 and $−382.1 billion in 2000) represented establishment of the U.S. unfavorable balance of trade.

Some secret U.S. Federal Reserve Bank loans added a burden and weighed heavily on the slow economic growth rate and the level of the U.S. economy. *Bloomberg Markets* magazine, November 27, 2011, in the article entitled "Secret Fed Loans Helped Bank Net $13 Billion" by Bob Ivry, Bradley Keoun, and Phil Kuntz[2] reported that the volume of money distributed and committed by the Federal Reserve Bank between 1985 and 2009 amounted to $7.77 trillion as of March 2009 for the purpose of the recovery of the financial and monetary system.

The U.S. gross domestic product (GDP) was $15,097.7 trillion for 2011, which represented just over a half larger in size than the volume of money distributed and committed by the Federal Reserve Bank between 1985 and 2009 that supported the recovery of the financial and monetary system. Globally, the authors of the article in *Bloomberg Markets* magazine,[3] Bob Ivry, Bradley Keoun, and Phil Kuntz, reported that the banks netted income estimated at $13 billion in earnings through the strategy of conducting below market rates transactions on emergency Federal Reserve Bank loans that covered the period August 2007 to April 2010.[4]

The preceding activities and practices of the Federal Reserve are in contrast to the Federal Reserve Act because the goals of monetary policy, according to the Federal Reserve Act, "specifies that the Board of Governors and the Federal Open Market Committee should seek to promote effectively the goals of maximum employment, stable prices and moderate long-term interest rates."[5]

It was former president Bill Clinton who recognized the urgency for the advancement and acceleration of the growth of the U.S. economy and who launched the pathbreaking, *National Nanotechnology Initiative: Leading to the Next Industrial Revolution.*[6] Nanotechnology, which is the lifeblood requirement for the U.S. nanotech industrialization–driven force and nanoscale source for the sustained and rapid acceleration in the magnitude and double-digit increase in the rate of growth of the U.S. economy, is powered by, and concerned with, materials and systems that are accompanied by inherent structures and components that display new and importantly advanced physical, chemical, and biological properties, phenomena, and processes based on their nanoscale size. Nanotechnology is the capability to operate, perform, and function at intra-atom by intra-atom (material nanoarchitectonics), molecular level, atom by atom, in the creation of large structures with fundamentally novel nanometer-scale organization, and seeing the nanoscale. The measurement in the definition of nanotechnology is on the nanometer scale or nanoscale range of 1 to 100 nanometers (nm)—where a typical dimension of 10 nm is 1,000 times smaller than the diameter of a human hair and 1 nanometer is equal to 1 billionth of a meter (10 to the negative minus 9). The behavior of structural features operates in the range of about 10 to the negative minus 9 to 10 to the negative minus 7.

President Barack Obama, in his ranking of national priorities for the U.S. economy, should select what the author's recommendation is for the number one priority, namely, the nanotechnology and nanoscale practical application global competitive advantage whose composition is bounded by the control of nanostructures, nanodevices, nanosystems, and nanomaterials at the atomic, intra-atom by intra-atom, and molecular levels while, at the same time, form the proliferation of networks of nanomanufacturing and nanofabrication—the necessary condition and competitive advantage requirement that would enable the acceleration of the U.S. economy and growth and diversification of the global economy. Within this context of the establishment of global competitive advantage economic performance in world trade relations, the description of a U.S. National Nanotechnology Initiative (NNI) Strategic Plan is research-based documentation. It is theoretical in its mission; it targets the essential centralization and mobilization of the Fortune 500 largest U.S. corporations, the critical state-sponsored industrial concentration, and the galvanization of nanotechnology practical application technologies that are required to be integrated in the relevant market base that includes the North American Industry Classification System (NAICS) United States 2012, Sector 31 to 33—Manufacturing; Sector 541711—Research and Development in Biotechnology; and Sector 22—Utilities.

THE GRAND STRATEGY

The creation, management, formation, and administration of a nanometer-scale practical application technologies–integrated industrialization public benefit corporation whose purpose would be (a) to establish the nanotechnology practical application technologies global competitive advantage and (b) to provide the origination of its own investment sources, development, formation, and financing of an operational nanotechnology accumulation and nanoscale integrated industrialization public benefit corporation, which by definition is the grand strategy; essentially, the aim shall be (c) to include the determination force that drives the vast transformation and convergence of nanotechnology practical application innovation and integration of the U.S. economy, and the huge size and rapid rate of increase, as well as the growth nanodiversification magnitude vector increase in U.S. gross domestic product A grand strategy is a plan of action so complete, it cannot be overcome by opponents' counteraction. Within this context of the North American Industry Classification System, the nanotechnology and nanometer-scale practical application technologies, particularly in connection with the preceding framework of the creation of a proposed nanotechnology and nanometer-scale practical application technology–embedded industrialization-formed, government-sponsored corporation act source, in effect, would constitute the relevant market characterization (at the same time including alternatives available to sellers) as presented in Table 1.1 a-d: Nanometer Scale Technology Driven Global Competitive Advantage. The nanotechnology phenomena, composition, and embodiment involve the revolutionary ability to manipulate matter, devices, materials, structures, and systems, which are, due to their nanoscale size and novel behavior where atoms, intra atoms' relations to intra atoms, and molecules are manipulated, in fact, are representations of discoveries and revolutionary breakthrough functions in new materials, properties, and systems.

It is important to note that the construction of Table 1.1 a-d derived from the collection, compilation, and interpretation of a substantial and vast array and variety of sources of Japanese nanotechnology and nanometer scale–driven production, distribution, and dissemination and the diffusion of concepts, issues, terminology, tools, press releases, and equipment that have been made available from numerous publicly available websites of corporations, MEXT (Japan's Ministry of Education, Culture, Sports, Science and Technology), National Institute of Materials Science's *NIMS NOW International* magazine–material nanoarchitectonics (http://www.nims.go.jp/eng/publicity/nimsnow/); *Japan Nanotech Bulletin*; *The Nikkei Weekly*; science, research and development, and technical journals of major Japanese corporations made publicly available on corporation websites.

This Table 1.1 a-d: Nanometer Scale Technology Driven Global Competitive Advantage, viewed from its revolutionary nanotechnology discoveries and advancement of nanometer-scale practical application, is the representation of the sources that, in essence, determine, drive, form, and generate significant changes and newly revealed phenomena, materials, and systems

Table 1.1a. Nanometer Scale Technology Driven Global Competitive Advantage

Nanosubstaces (Multifuction)	Nanocrystals
Nanoteh practical application	Nanofilms
NanotechJapan	Nano-porous titanate
Nanotech network program Japan	Nano-green field
Nanostructures	Nanomagnets
Nanodevices	Nanogranular film
Nanomaterials	Nanowires
Nanomanufacturing	NanoRegions magnetic field
Nanoparticles	Nanolinks
Nanobiomaterials	Epoxy/mica nanocomposite
Nanochemistry	Nanostructured ferroelectrics
Nanofabriction	Nanometer thickness
Nanofibers	Nanocube
Nanoevaluation technologies	Nanocharacterization
Controlled nanoscale doping	Nanotechnology encryption
High-k nanodielectrics devices	Nanoprobe
Nanoelectronics devices	Nanomeasurements
Titano-niobates nanosheets	Nanowire

Table 1.1b. Nanometer Scale Technology Driven Global Competitive Advantage

Nanotechnology-driven materials	Nanododecahedron
Nanotechnology network	Nanohetero core shell
Nanomaterials science	Nanoelectronics
Nanosystems	Nanotubes
Nanoballs	Nanostrands
Nanoprocessing	Nnoceramics
Nanofiundry	Nanomechstronics
Nanoscience	Nanocone array
Nanofibrillar materials	Naosystems
Nanoengineering	Nanometer scale
Nanobiotechnology	Nanoionics
Nanotrac wave	Nanoelectromechanical
Nanoimprint technology	Nanolines
Carbon nannohorn	Graphene nanoplatelets
Materials nanoarchitectonics	Nanophotonics
Nanoporous Au catalyst	Nanocavity

Table 1.1c. Nanometer Scale Technology Driven Global Competitive Advantage

Nano optical elements	Single wall carbon nanotubes (SWCNT)
Metal nanopatterned surface	Nano superconductivity
Nanosphere lithography	Nano shells
Nano schottky diodes	Au nanorod
Nanoscopic observation	Nanoplasmonic structures
Nano appertures	Nano thermometer using carbon nanotubes
Nanometers	Dimension of ten nanometers (nm) is 1,000 times smaller than the diameter of a human hair
Nano enabled	Nano carrier for delivery of DNA vaccine into immune cells
Nanosecond	Nano antennas
Nanodomain structure	Nano manipulation
Nanodesign	Nano brain
Nanotechnology platform JAPAN	Nano oriented crystal
Nanoarticles	Nano innovation
Nanoclusters	Nano means 1 billionth

Table 1.1d. Nanometer Scale Technology Driven Global Competitive Advantage

One nanometer (nm) is 1 billionth of a meter	Bionanotubes
An atom is about 0.1 nm to 0.3 nm in size	Nanofunctionality
Organic nanotubes	Nanotesters
Graphene nanotubes	Nano interface of living tissue

whose devices, components, and structures include novel and improved physical, chemical, and biological properties and processes, in fact, are inherent and intrinsic to their nanometer-scale generation and implementation of productivity and economic growth.

The nanotechnology practical application and innovation and the advancement of knowledge at nanoscale not only produce revolutionary nanotech production techniques but contribute new management decision-making methods and new forms of business organization, such as Nanotechnology Platform Japan, nanotechnology keiretsu platform, and the nanotechnology multinational enterprise platform.

For example, the following three listed resource areas, whose purpose is to serve as the foundation for the creation of nanometer-scale (one nanometer is one-billionth of a meter) technology drivers that would enable the diffusion, generation, and multiplier effect of Table 1.1 a-d: Nanometer Scale Technology Driven Global Competitive Advantage to be a necessity in order to propel aggregate economic growth: (a) the U.S. National Income and Product Accounts (NIPA), Bureau of Economic Analysis, U.S. Department of Commerce; (b) U.S. Federal Reserve Board Index of Industrial Production Activity; and (c) the North American Industry Classification System.

First, to strengthen the sources of aggregate economic growth, while at the same time ensuring the acceleration of sustainable economic growth, it is essential that nanotechnology mobilizes as a base for the creation and formation of nanometer-scale application drivers enabling the integration necessary to propel persistent and sustained increases in the National Income and Product Accounts (NIPA), especially the Gross Domestic Product (GDP), which is the output of goods and services produced by labor and property located in the United States, or the market value of all final goods and services produced in the economy in one year.

More specifically, the integration of nanometer-scale technology (one nanometer is one- billionth of a meter) would add novel and revolutionary nanoscale value, innovation, and nano- replacements technology to the accumulation of stocks of private plants, machinery and equipment, government fixed assets, and inventories owned by private business enterprises. The fixed assets, which are defined in the concepts and methods of the U.S. National Income and Product Accounts, "are produced assets that are used repeatedly, or continuously, in the processes of production for more than 1 year. The Bureau of Economic Analysis (BEA) estimates cover structure, equipment and software, but not cultivated assets such as livestock or orchards. The acquisition of fixed assets by private business is included in the NIPA component gross private domestic investment, and the acquisition of fixed assets by government is included in the NIPA measure, government consumption expenditures and gross investment. The depreciation of fixed assets—that is, the decline in their value due to wear and tear, obsolescence, accidental damage, and aging is captured in the NIPA measure, consumption of fixed capital."[7]

There would be the creation of a powerful set of nanometer-scale technology tools aimed at aggregate economic growth maximization if the nanometer-scale technology diffusion, advancement, and its multiplier effect were to dominate the two components of Gross Domestic Product, namely, gross private domestic investment and government consumption expenditures and gross investment.

The dollar value of the gross private domestic investment component of GDP, which reflects a propelling factor and dynamic source in the progression and increase in economic growth, has developed a pattern that illustrates a disturbing trend. For example, the gross private domestic investment component of GDP was $1,795.2 billion (2011); $1,714.9 billion (2010); $1,454.2 billion (2009); $1,939.9 billion (2008); $2,159.5 billion (2007); and $2,231.8 billion (2006), and as a result, this pattern is clearly one of declining investment and slowdown in economic growth.[8]

The dollar value of the government consumption expenditures and gross investment component of Gross Domestic Product is the representation of a significant factor in the determination of economic growth, and it demonstrated a downward trend pattern. For example, the government consumption expenditures and gross investment component of GDP was $3,029.7 billion (2011); $3,002.8 billion (2010); $2,917.5 billion (2009); $2,878.1 billion (2008); $2,674.2 billion (2007); and $2,518.4 billion (2006).[9]

The joint components, (1) gross private domestic investment component of GDP and (2) government consumption expenditures and gross investment component of GDP, have presented a combination of enormous financial stabilization sources that strengthen capital formation and economic growth.

Essential to the measurement of the gross private domestic investment component of GDP assessment are the significant nanotechnology contribution and the nanometer-scale technology innovation, integration, and replacement of "Less: Consumption of Fixed Capital, Total" (or the economic charge for the using up of private and government fixed capital located in the United

States), which is defined as the decline in the value of the stock of assets due to wear and tear, obsolescence, accidental damage, and aging, and is reported quantitatively as $1,950.0 billion (2011); $1,874.9 billion (2010); $1,866.2 billion (2009); $1,854.1 billion (2008); $1,767.5 billion (2007); $1,660.7 billion (2006).[10] The "Less: Consumption of Fixed Capital, Total" was extracted from Table B-26, *Economic Report of the President Transmitted to the Congress February 2012*.[11]

Second, the current Federal Reserve Board (FRB) monthly Index of Industrial Production Activity is essentially outdated, indicating that a replacement is necessary. If the creation of nanometer-scale application technology full utilization, Table Alpha 01: Nanometer Scale Technology Driven Global Competitive Advantage, were integrated in the composition of the FRB index, the result would be nanoscale innovation and novel phenomena advantages and achievements that arise from the contribution that is provided through the revolutionary changes, radical improvements, added value, and supraproductivity efficiency to be delivered as a result of revolutionary nanotechnology practical application.

The FRB Index of Industrial Production Activity covers manufacturing, mining, and electric and gas utilities, and it measures the real output of the manufacturing, mining, and electric and gas utilities industries; the reference base period for the index is 2007, according to the FRB. The manufacturing sector consists of those industries that are included in the North American Industry Classification System (NAICS).[12]

An integration of the pervasiveness and diffusion of nanometer-scale technologies application, Table Alpha 01, in the composition base of the Federal Reserve Board Index of Industrial Production Activity would have the effect of making a substantial contribution to the total performance of the new, nanotechnology-driven Federal Reserve Board Index of Industrial Production Activity, which measures real output and helps illuminate structural developments in the economy.[13]

Third, the NAICS[14] resource area, whose purpose is to serve as a base for the creation of nanometer-scale technology drivers, would therefore activate the multiplier effect to be in operation and, thereby, propel the advancement, progression, and sustained increase in the rate, level, and size of aggregate economic growth–based nanotechnology.

The U.S. economy is divided into twenty sectors by the *NAICS United States 2012 Manual* and included in these twenty sectors are industries that are classified and based on the production criterion in which four sectors represent goods-producing industries and sixteen sectors represent service-producing industries. There are 1,175 industries in the NAICS covering all economic activities. Moreover, NAICS is the presentation and comprehensive description of industries in the United States. For example, NAICS is designed for statistical purposes and is used by U.S. statistical agencies to facilitate the collection, tabulation, presentation, and analysis of data relating to establishments, and to provide uniformity and comparability in the presentation of statistical data. NAICS is an industry classification system that groups establishments into industries based on the similarity of their production processes.

GUIDELINES SOURCES FOR CREATION OF DESIGN/FRAMEWORK FORMATION TYPES OF TABLE 1.1 A-D: NANOMETER SCALE DRIVEN GLOBAL COMPETITIVE ADVANTAGE GOVERNMENT SPONSORED ENTERPRISES

In the gauging of the significance, institutional dynamics, policy, and the act of Congress that have created government-sponsored enterprises, it is the author's view that the following selections and historical precedents, in fact, form the representation of a framework and set of guidelines that are essential in the formulation of a nanotechnology-driven grand strategy.

A. War Finance Corporation Act, Approved April 5, 1918, (40 Stat. 506), Guaranty Trust Company of New York, 140 Broadway, New York, NY: "An Act to provide further for the national security and defense and, for the purpose of assisting in the prosecution of the war, to provide credits for industries and enterprises in the United States necessary or contributory to the prosecution of the war, and to supervise the issuance of securities, and for other purposes. Be it enacted by the Senate and House of Representatives of the United States of America in Congress assembled" (http://www.archive.org/deetails/warfinancecorpor00guarrich).
B. Reconstruction Finance Corporation Act as amended and Provisions of the Emergency Relief and Construction Act of 1932, and the provisions of Title I and Title II of the latter act pertaining to the Reconstruction Finance Corporation, July 21, 1932, Washington, DC: U.S. Government Printing Office, 1932. The Emergency Relief and Construction Act of 1932; an act to provide emergency financing facilities for financial institutions, to aid in financing agriculture, commerce and industry, and for other purposes.
C. Peace Treaty of Versailles, Article 231-247 and Annexes, Reparations; Part VIII, Reparations, Section 1, General Provisions (http://www.net.lib.byu.edu/-rdh7/wwi/versa/versa7.hyml).

D. James C. Bonbright and Gardner C. Means, *The Holding Company: Its Public Significance and Its Regulation*, 1st ed., New York: McGraw-Hill Book Company, Inc., 1932 (reprint ed., Augustus M. Kelley Publishers, New York, NY, 1969).
E. Kevin R. Kosar, analyst in American National Government, Government and Finance Division, Congressional Research Service–the Library of Congress, CRS Report for Congress: "Government-Sponsored Enterprises (GSEs): An Institutional Overview," prepared for members and committees of Congress, Order Code RS21663, updated April 23, 2007. Congress charted government-sponsored enterprise (GSEs) to improve the workings on credit markets. This report briefly describes the nature of GSEs, their mixed governmental private nature, the differences between GSEs and government agencies, and concerns about and supporting arguments for GSEs. This report will be updated as events warrant.
F. New York State Urban Development Corporation, 2006 New York Code–Urban Development Corporation Act, Chapter 174 of the laws of 1968: "It is hereby declared to be the policy of the State to promote a vigorous and growing economy, to prevent economic stagnation and to encourage the creation of new job opportunities in order to protect against the hazards of unemployment, reduce the level of public assistance to now indigent individuals and families, increase revenue to the State and to its municipalities, and to achieve stable and diversified local economies."
G. Jon Gertner, *The Idea Factory: Bell Labs and the Great Age of American Innovation*, New York: Penguin Group (USA), 2013. A sweeping atmospheric history of Bell Labs that highlights its unparalleled role as an incubator of innovation and birthplace of the century's most influential technologies. AT&T Bell Laboratories, at its peak, employed nearly fifteen thousand people, twelve hundred of whom had PhD degrees. Thirteen would go on to win Nobel Prizes. Bell Labs was, in effect, a factory of ideas whose workings have remained largely hidden until *The Idea Factory*.

Among countries immersed in the establishment and implementation of economic growth as a national priority, this economic growth (Table 1.1 a-d: Nanometer Scale Technology Driven Global Competitive Advantage) practical application surpasses, in essence, all previous determinants of economic growth, thereby, the global competitive advantage.

NOTES

1. *Economic Report of the President Transmitted to the Congress February 2012, Together with the Annual Report of the Council of Economic Advisors*, U.S. Government Printing Office, Washington, DC, 2012, Table B-211: Gross Domestic Product, 1963–2011, p. 316, Appendix B.

2. Executive Office of the President, Office of Management and Budget, North American Industry Classification System, United States, 2007, Sections 22 and 31-33; *United Nations International Standard Industrial Classification of All Economic Activities*, rev. 4.

3. *Bloomberg Markets*, "Secret Fed Loans Helped Banks Net $13 Billion," November 28, 2011, p. 2 (http://www.bloomberg.com/neews/print/2011-11-28/secret-fed-loans-undisclosed-to-congr . . . 11/28/2011).

4. Ibid., p.1.

5. Board of Governors of the Federal Reserve System, *The Federal Reserve System Purposes & Functions*, Washington, DC, June 2005, passim

6. *National Nanotechnology Initiative: Leading to the Next Industrial Revolution*, a report by the Interagency Working Group on Nanoscience, Engineering and Technology (IWGN), Committee on Technology, National Science and Technology Council, Washington, DC, February 2000; National Institute for Materials Science, "Nano-Bio Medical Science Meets Nanoarchitectonics," *NIMS NOW International*, no. 4, May 2012, passim; Harris, James A., Sr., *Nanotechnology World Economy*, Bloomington, IN: AuthorHouse, 2009, pp. 3–6; National Institute for Materials Science, "Seeing the Nanoworld NIMS Transmission Electron Microscopy Station," *NIMS NOW International*, no. 5, June 2013; National Science and Technology Council Committee on Technology Subcommittee on Nanoscale Science, Engineering, and Technology, *National Nanotechnology Initiative Strategic Plan*, Washington, DC: Executive Office of the President, National Science and Technology Council, February 2011, passim.

7. Bureau of Economic Analysis, *Concepts and Methods of the U.S. National Income and Product Accounts*, chapters 1–7, Washington, DC, November 2010, pp. 2–3.

8. *Economic Report of the President Transmitted to the Congress February 2012 Together with the Annual Report of the Council of Economic Advisors*, Washington, DC: U.S. Government Printing Office, 2012; Table B-1: Gross Domestic Product, 1963–2011, p. 316, appendix B.

9. Ibid., p. 317.

10. Ibid., Table B-26, p. 348.

11. Ibid.

12. Board of Governors of the Federal Reserve System, "Industrial Production and Capacity Utilization–G.17," 2012 (http://www.federalreserve.gov/releases/g17/about.htm).

13. Board of Governors of the Federal Reserve System, "Industrial Production and Capacity Utilization–G.17," Industrial Production Explanatory Notes, 2012 (http://www.federalreserve.gov/releases/G17/I.Notes.htm).

14. Executive Office of the President, Office of Management and Budget, "North American Industrial Classification System United States, 2007," passim.

2

Definitions and Concepts of Global Over-the-Counter Derivatives Accumulation Crises

The central bank members of the Bank for International Settlements (BIS) have contributed substantially to the continued growth in the size, structure, and complexity of global over-the-counter (OTC) derivatives transactions as financial instruments in the composition of the financial system. In the unraveling of the distinction and complexities of OTC derivatives the following set of definitions and concepts that are presented will add some clarity while limiting or reducing complexity.

A. U.S. Commodity Futures Trading Commission (CFTC).[1]

The CFTC, in its glossary, has defined over-the-counter (OTC) as trading of commodities, contracts, or other instruments not listed on any exchange. The OTC transaction can occur electronically or over the telephone, which is also referred to as off-exchange.

The CFTC definition of derivatives is a financial instrument, traded on or off an exchange, the price of which is directly dependent upon (i.e., "derived from") the value of one or more underlying securities, equity indices, debt instruments, commodities, other derivative instruments, or any agreed upon pricing index of arrangement (e.g., the movement over time of the Consumer Price Index or freight rates).

Derivatives are used to hedge risk or to exchange a floating rate of return for a fixed rate of return. Derivatives include futures, options, and swaps. For example, futures contracts are derivatives of the physical contract and options on futures are derivatives of future contracts.

B. "Global Over-the-Counter Derivatives Markets and the Commodity Exchange Act Report of the President's Working Group on Financial Markets."[2]

"In this report of the President's Working Group on Financial Markets, the over-the-counter (OTC) derivatives instruments importance in the financial system is uncovered through the market for global OTC derivatives that has expanded steadily and rapidly over the past two decades. The trend shows a pattern and direction; for example, at year-end 1998, the total estimated notional amount (reference to a principal base) of outstanding global OTC derivatives contracts was $80 trillion, reflecting an increase of 11 percent from June 1998 as reported by the Bank for International Settlements (BIS). The vast majority of global OTC derivatives at year-end 1998 were interest rate and foreign exchange contracts, 72 percent and 26 percent respectively, according to BIS; the equity-related contracts made up only 2 percent of the market. The President's Working Group states that the activity in global OTC derivatives markets was primarily concentrated in three types of instruments: swaps agreements, options, and hybrid instruments.

C. U.S. Derivatives Contracts with Foreigners, Holdings and Transactions, U.S. Department of the Treasury.[3]

The U.S. Department of the Treasury Resource Center, taking action in response to a definition of derivatives, reported that a derivatives contract is a financial contract whose value is derived from the values of one or more underlying assets, reference rates, or indexes of asset values or reference rates. Moreover, the common types of derivatives contracts include forwards, futures, swaps, and options. The Treasury International Capital Reporting System (TIC) collects data on derivatives contracts that meet the Financial Accounting Standards Board (FASB) Statement No. 133 definition.[4]

Stephanie E. Curcuru, Division of International Finance, Federal Reserve Board, has expressed the view that the global derivatives market has grown and expanded rapidly in the past decade, while making it known that by one measure of market size—the notional value, which is used to determine the payment made on derivatives contracts—the derivatives market expanded from $87 trillion in June 1998 to the huge level of $454 trillion in June 2006.

Stephanie E. Curcuru, in her prepared statement released under the title "Types of Derivatives" included the following assessment: "A derivative is a financial contract whose value is derived from something else (such as the value of a stock or bond), referred to as the 'underlying.' Anything that can be measured can serve as the underlying to a derivative. The underlying can be the price of a stock, the yield of a bond, a credit rating, the value of an index, or even something more exotic, like the average temperature in a region over a given period. Common types of derivatives include options, forwards, futures, and swaps. In an options contract, the buyer pays an up-front premium for the right, but not the obligation, to purchase or sell a specified quantity of the underlying at specified price on (or, in some contracts, until) the expiration date. The two parties to a forward or futures contract agree to exchange assets or their cash equivalent at, or until, a future date. The two parties to a swap agree to exchange assets or their cash equivalent periodically until a future date."[5]

D. U. S. Cross-Border Derivatives Data: A User's Guide by Stephanie E. Curcuru, Federal Reserve Bulletin, May 2007[4]

Stephanie E. Curcuru, Division of International Finance, Federal Reserve Board, has expressed the view that the global derivatives market has grown and expanded rapidly in the past decade, while making it known that by one measure of market size—the notional value, which is used to determine the payment made on derivative's contract—the derivatives market expanded from an amount of $87 Trillion in June 1998 to the huge level of $454 Trillion in June 20006.

Stephanie E. Curcuru, in her prepared statement released under the title, Types of Derivatives, included in the following assessment, "A derivative is a financial contract whose value is derived from something else (such as the value of a stock or bond), referred to as the 'underlying.' Anything that can be measured can serve as the underlying to a derivative. The underlying can be the price of a stock, the yield of a bond, a credit rating, the value of an index, or even something more exotic, like the average temperature in a region over a period. Common types of derivatives include options, forwards, futures, and swaps. In an option contract, the buyer pays an up-front premium for the right, but not the obligation, to purchase or sell a specified quantity of the underlying at specified price on (or, in some contracts, until) the expiration date. The two parties to a forward or futures contract agree to exchange assests or their cash equivalent at, or until, a future day. The two parties to a swap agree to exchange assets or their cash equivalent periodically until a future date.[5]

E. Keynote address of Chairman Gary Gensler (CFTC), "OTC Derivatives Reform, Market's Outlook," OTC Derivatives Markets Conference, March 9, 2013.[6]

Chairman Gary Gensler, U.S. Commodity Futures Trading Commission (CFTC), in highlighting the over-the-counter derivatives reform, remarked that a comprehensive regulatory framework governing the over-the-counter derivatives should apply to all dealers and all derivatives, no matter where traded or marketed. It should include interest rate swaps, currency swaps, foreign exchange swaps, commodity swaps, equity swaps, credit default swaps, and any new product that might be developed in the future. Chairman Gensler also stated that effective reform of the marketplace requires three critical components. First, the derivatives must be explicitly regulated; second, to promote public transparency, standard over-the-counter derivatives should be traded on exchanges or other trading platforms; and third, standard OTC derivatives should be brought to clearinghouses in order to lower risk further. In his assessment of the OTC derivatives markets reform, Chairman Gensler (CFTC) singled out credit default swaps (CDS) products that directly contributed to the financial crisis of 2007–2008. The credit default swap is a bilateral over-the-counter contract in which the seller agrees to make a payment to the buyer in the event of a specified credit event in exchange for a fixed payment or series of fixed payments. The market for credit default swaps has grown exponentially within the last decade, according to Chairman Gensler, who also made it clear that the marketplace has grown from a notional value of an estimated $630 billion in the second half of 2001 to a huge level of $36 trillion by the end of 2009. In addition, Chairman Gensler continued, the Bank for International Settlement (BIS) data show that more than 95 percent of credit defaults swaps transactions are between financial institutions.

F. Federal Reserve Bank of New York (FRBNY) Staff Report, *An Analysis of OTC Interest Rate Derivatives Transaction: Implications for Public Reporting,* Staff Report No. 557 by Michael Fleming, John Jackson, Ada Li, Asani Sarkar, and Patricia Zobel, March 2012.[7]

The authors, M. Fleming, J. Jackson, A. Li, A. Sarkar, and P. Zobel, have made a substantial contribution through the magnitude and the diversity in the assessment of the transaction data from the OTC interest rate derivatives (IRD) market's broad scope with a wide range of products, currencies, and maturities traded. The question arises, what is your explanation for the cause of the financial crisis of 2007–2008; has the financial crisis ended? The authors of the FRBNY Staff Report No. 557 have made an excellent presentation concerning the definition of derivatives. It is their view that a derivative is a financial instrument whose value depends upon that of another asset. A derivative may be used as a tool to either take a position on the underlying asset or to transfer or hedge risk. Derivatives can either be traded on organized exchanges or negotiated privately between two parties. Privately negotiated trades, known as over-the-counter or OTC trades, allow parties to customize features of the derivative to serve the specific needs of the users.

An interest rate derivative (IRD) is an agreement to exchange payments based on different rates over a specified period of time. In its most common form, the single currency interest rate swap, the parties agree to exchange payments periodically based on a fixed interest rate agreed upon at the outset of the transaction and a floating interest rate based on a specified reference index. The U.S. dollar interest rate swaps typically reference the three-monnth LIBOR (LondonInterBank Offered Rate) index, and participants usually pay the floating payments at three-month intervals and fixed payments at six-month intervals over the life of the contract, the authors of the FRBNY Staff Report No. 557 noted.

G. Testimony: Patrick M. Parkinson, Deputy Director, Division of Research and Statistics, "Over-the-Counter Derivatives," before the Subcommittee on Securities, Insurance, and Investment, Committee on Banking, Housing, and Urban Affairs, U.S. Senate, July 2008.[8]

A credit derivative, Patrick M. Parkinson stated in addressing the issue of "Over-the-Counter Derivatives," is a financial contract whose value is derived from the value of debt obligations issued by one or more reference entities. The predominant type of credit derivative is a credit default swap (CDS). Patrick M. Parkinson, in the delivery of his testimony, explained that in a credit default swap a "protection buyer" pays premiums to a "protection seller." In return, and in the event of a default or other specified credit event, the protection seller is obligated to pay the protection buyer the notional or par value for the debt, thereby transferring the risk of default from the buyer to the seller.

Of particular importance regarding the risks of using credit derivatives is counterparty credit risk—that is, the risk that a counterparty to a credit derivatives contract could fail to perform its contractual obligations, resulting in losses to the non-defaulting counterparty. Therefore, Patrick M. Parkinson continued, in the case of a credit default swap, for example, if the protection seller itself should become insolvent, the protection buyer would lose the value of that protection and would need to replace it by purchasing protection from another seller. If the premiums required by the market for protection against default by the reference entity had risen since the protection had been purchased from the insolvent seller, then the protection buyer would be exposed to a loss that is equal to the present value of the difference between the premiums paid on the new contract and the premiums paid on the original contract.

H. Lawrence H. Summers, Secretary, U.S. Department of the Treasury; Alan Greenspan, Chairman of the Board of Governors of the U.S. Federal Reserve System; Arthur Levitt, Chairman of the U.S. Securities and Exchange Commission; and William J. Rainer, Chairman of the U.S. Commodity Futures Trading Commission, *Over-the-Counter (OTC) Derivatives Markets and the Commodity Exchange Act,* Report of the President's Working Group on Financial Markets, November 1999.[9]

The Report of the President's Working Group on Financial Markets, "Over-the-Counter (OTC) Derivatives Markets and the Commodity Exchange Act," has formed the assessment that over-the-counter derivatives have transformed the world of finance, made increases in the range of financial products that are available to corporations and investors, and fostered more precise ways of understanding, quantifying, and managing risks. At the end of 1998, the Report of the President's Working Group (hereinafter referred to as the Working Group) has acknowledged that the estimated notional value of OTC derivatives contracts was $80 trillion. The trading of financial derivatives by eligible swap participants should be excluded from the Commodity Exchange Act (CEA). To do otherwise, the "Working Group" report continued, would perpetuate legal uncertainty or impose unnecessary regulatory burdens and constraints upon the development of these markets in the United States. Thus, the Working Group, among its conclusions, noted that it is significant to remove legal impediments to the development of electronic trading systems. More particularly, and with reference to OTC derivatives, the Working Group listed among its recommendations the unanimously chosen decision, which is the exclusion from the Commodity Exchange Act for electronic trading systems for derivatives provided that the trading systems limit participation to sophisticated counterparties trading for their own accounts and are not used to trade contracts that involve non-financial commodities with finite supplies.

The Working Group, through its examination of the issue, "Enhancing Legal Certainty Swaps," in fact, agreed that there is no compelling evidence of problems involving bilateral swap agreements that would warrant regulation under the CEA. Therefore, the Working Group made the following unanimous recommendation: To the extent that OTC derivatives transactions between eligible swap participants are excluded from the CEA, they should also be excluded from coverage under certain state laws (such as laws designed to regulate gambling or bucket shops) that might be construed to prohibit or inappropriately regulate such transactions.

The Working Group members have agreed that the electronic trading systems for OTC derivatives have only just begun to emerge on a widespread basis, and such electronic trading systems should be allowed to grow, unburdened by a new anticipatory statutory structure that could prove entirely inappropriate to their eventual evolution. Herein, the Working Group laid the foundation, framework, and non-regulatory governance conditions and operation causes of the economic and financial collapse of 2007–2008.

I. Testimony: Kevin Warsh, Board of Governors of the U. S. Federal Reserve System, "Hedge Funds," before the Committee on Financial Services, House of Representatives, July 11, 2007.[10]

Federal Reserve Board (FRB) Governor Kevin Warsh's testimony included the position that the Board of Governors of the Federal Reserve System advocated that the increased scale and scope of hedge funds has brought important net benefits to financial markets and have the potential to reduce systemic risk. This net benefit is achieved by the dispersion of risks more broadly and by serving as a large pool of opportunistic capital that could stabilize financial markets in the event of disturbances. After a search of the literature, there has not been an exact or precise legal definition of hedge funds, according to FRB Governor Kevin Warsh, who continued by noting that the term hedge funds generally refers to a pooled investment vehicle that is privately organized, administered by a professional investment manager, and not widely available to the public. The emphasis placed on market discipline, in fact, is neither an endorsement of the status quo nor does it imply a passive role for government. It was the position of FRB Governor Warsh that, in recent years, the global banks have significantly strengthened their practices and procedures for the management of risk exposures to hedge funds. In this financial field greater advancement is required because of the increasing complexity of structured credit products such as collateralized debt obligations (CDOs), which are securities that entitle the purchaser to some portion of the cash flows from a portfolio of assets that may include bonds, loans, mortgage-backed securities, or other CDOs. For a given pool or portfolio, the CDOs issued may be designated as senior debt, mezzanine debt, subordinated debt, and equity. There are some valuable insights referenced in the testimony of FRB Governor Warsh, such as his assessment that the key creditors and counterparties of the hedge are, in fact, a relatively small group of global commercial and investment banks. These global banks provide credit to hedge funds through securities repurchase agreements (repos) and acting as counterparties to the hedge funds' global over-the-counter derivatives and exchange-traded derivatives, in effect, the involvement of financial practices that were inherent in the Wall Street economic and financial collapse in 2007–2008.

J. Carl Levin, Chairman, and Tom Coburn, Ranking Minority Member, U.S. Senate Permanent Subcommittee on Investigations, Committee on Homeland Security and Governmental Affairs, *Wall Sreet and the Financial Crisis: Anatomy of a Financial Collapse*, Majority and Minority Staff Report, April 13, 2011.[11]

This report by the U.S. Senate Permanent Subcommittee on Investigations, hereinafter referred to as "This Report," is representative of the four subcommittee hearings that were held during the April 2010 examination of four root causes of the financial crisis in 2007–2008. The subcommittee presentation examined the evidence that showed how high-risk lending by financial institutions, the regulatory failure, inflated credit ratings, and the high-risk and poor-quality financial products designed and sold by some investment banks, in fact, contributed to the financial crisis.

This Report, through its examination of how investment banks contributed to the financial crisis, uncovered how large investment banks designed and promoted complex financial instruments, which are often referred to as structured finance products, that were at the heart of the crisis in 2007–2008. They included residential mortgage-backed securities (RMBS) and collateralized debt obligations (CDO) securities, credit default swaps (CDS), and CDS contracts that are linked to the ABX index. These complex, high- risk financial products, according to This Report, were engineered, sold, and traded by the major investment banks.[12]

The Permanent Subcommittee on Investigations has found that from 2004 to 2008, U.S. financial institutions issued nearly $2.5 trillion in residential mortgage-backed securities (RMBS) and issued over $1.4 trillion in CDO securities that were backed primarily by mortgage-related products. While investment banks typically charge fees of $1 million to $8 million to act as the underwriter of an RMBS securitization, and $5 million to $10 million to act as the placement agent for a CDO securitization.[13]

In its focus on derivatives, the Permanent Subcommittee on Investigations reported results that indicate some investment banks modified still another structured finance product, a derivative known as a credit default swap (CDS), for use in the mortgage market. It is similar to an insurance contract. For instance, a credit default swap is a contract between two parties in which one party guarantees payment to the other if the assets referenced in the contract lose value or experience a negative credit event. The party selling the insurance is referred to as the "long" party, since it profits if the referenced asset should perform well. The party buying the insurance protection is referred to as the "short" party because it profits if the referenced asset should perform poorly.[14]

It is significant to know that credit default swaps contracts that reference a single, specific security or bond for protection against a loss in value or negative credit event have become known as "single name" RMBS or CDO security. The key significance of the credit default swap product for the mortgage market was that it offered an alternative to investing in residential mortgage-backed securities (RMBS) and collateralized debt obligations (CDO) securities that would perform well. The "single name" CDS contract would pay off if specific RMBS or CDO securities lost value or failed.[15]

K. Central Bank of Japan, "The Bank of Japan's Initiatives on the Macroprudent Front,"

October 18, 2011.[16]

In challenging and overcoming events such as the failure of Lehman Brothers Holdings in the United States, the Bank of Japan had concluded that it has become globally understood that the prevention of a financial crisis from re-emerging involves the analysis and assessment of risks in the entire financial system and adequate measures to prevent systemic risk from materializing. These findings, the Central Bank of Japan reported, define the so-called emphasis on the macroprudential framework.

There is a global attraction to the macroprudential framework especially since the recent global financial crisis triggered by the failure of Lehman Brothers Holdings. The lesson learned, the Central Bank of Japan noted, is that behind this trend exists a growing perception that the stability of the entire financial system is not necessarily achieved solely by ensuring the financial soundness of individual financial institutions.[17] In other words, the statement delivered by the Central Bank of Japan was strong: Risks that are supposed to be contained at individual financial institutions could in the aggregate destabilize the entire financial system. In addition, the spread of new financial technologies such as derivatives and the rise of new types of investors such as hedge funds, in fact, have made it difficult to identify risks in the financial system. Before the global financial crisis struck in the U.S. financial markets in 2007–2008, for example, the banks actively traded securitized products and credit derivatives with various institutional investors and financial institutions in order to transfer credit risk.[18]

The initiatives of the Central Bank of Japan reinforce the characteristics of its central bank performance in taking various initiatives on the macroprudential front. These types of macroprudential initiatives included (i) analysis and assessment of the financial system, i.e., the financial system is the embodiment of financial institutions, financial markets, and financial infrastructure such as payment and settlement systems and stability; (ii) better assessments of the robustness that use macro stress testing; (iii) assessments of financial imbalances that use macro indicators; (iv) identification of the risks that are observed in financial markets; and (v) coordination with microprudential activity of on-site examination and off-site monitoring.[19]

L. (Full Text) Commodity Futures Trading Commission, 17 CFR Part 1, RIN 3038-AD06; Securities and Exchange Commission, 17 CFR Part 240, Release No. 34-66868, File No. 87-39-10; RIN 3235-AK65, "Further Definition of 'SWAP Dealer,' 'Security-Based SWAP Dealer,' 'Major SWAP Participant,' 'Major Security-Based SWAP Participant,' and 'Eligible Contract Participant,' Action: Joint Final Rule; Joint Interim Final Rule; Interpretations. U.S. Securities and Exchange Commission jointly with U.S. Commodity Futures Trading Commission (CFTC) press release; "SEC Adopts Rule Defining SWAPS- Related Terms for Regulating Derivatives," April 18, 2012.[20]

This press release reported that the Securities and Exchange Commission (SEC) has unanimously adopted a new rule to define a series of terms that related to the over-the-counter (OTC) swaps market. The press release included the declaration that the rules, which were written jointly with the CFTC, have the impact of implementation of provisions of the Dodd-Frank Wall Street Reform and Consumer Protection Act (Dodd-Frank) that established a comprehensive framework for regulating derivatives.

Title VII of Dodd-Frank authorizes the SEC to regulate security-based swap dealers and major security-based participants and create a system by which they could register with the SEC. The press release highlights Title VII of the Dodd-Frank Wall Street Reform and Consumer Protection Act which defines the relevant terms and directs the SEC and the CFTC jointly to further define those terms in consultation with the Board of Governors of the Federal Reserve System. Those terms include "SWAP Dealer," "Security-Based SWAP Dealer," "Major SWAP Participant," "Major Security-Based SWAP Participant," and "Eligible Contract Participant."[21]

The press release from the SEC addresses the controversial de minimis exception, which is that the Dodd-Frank Act directs the SEC to implement a de minimis exception from the "security-based SWAP dealer" definition for a person who "engages in a de minimis quantity of security-based SWAP dealing. . . ." It directs the SEC to establish factors for determining when someone falls within this exception.

In contrast with the de minimis exception there are some watchdog groups that believe regulators are forming an important escape route that will establish some major problems. The de minimis exception or exemption, the *New York Times* has reported, would excuse companies from having to post additional capital and file reports. The controversy resides in this de minimis exemption which the *New York Times* described as a sort of regulatory hall pass for those companies that possess the insignificant derivatives holdings of $8 billion.[22]

M. The Banco Central do Brasil (Central Bank of Brazil), federal agency member of the National Financial System.[23]

The Brazilian Payment System, Securities, Derivatives, and Foreign Exchange Settlement Systems, through its press release— (http://www.bcb.gov.br/?intosecSecurities)—announced that in Brazil nearly 100 percent of all securities are dematerialized. In other words, the trading, clearing, and settlement systems are highly automated, and straight-through processing (STP) is broadly used.

In a separate press release, BM&FBOVESPA Derivatives, the spot market, forward, futures, options and swaps contracts are traded at the BM&FBOVESPA Derivatives Clearinghouse. These trades are mostly referred to and include interest rates, foreign exchange rates, and price and stock indexes. In addition, the related obligations are settled on trade date (T) plus one day, at the Banco Central do Brasil. The BM&FBOVESPA Clearinghouse acts as central counterparty and guarantees the settlement of Clearinghouse members' multilateral net positions. The Clearinghouse registers OTC derivatives whose settlement can or cannot be guaranteed according to parties' option. In the derivatives market, the BM&FBOVESPA Clearinghouse processes the daily margin calls, thereby, always marking to market the collateral and applying a haircut.[24]

N. The Central Bank of the Russian Federation (Bank of Russia), Moscow, Russia[25] (http://www.cbr.ru/eng/tody/default.aspx).

In a press release, the Central Bank of the Russian Federation (Bank of Russia) announced that its monetary policy was designed to maintain financial stability and create conditions that are conducive to sustainable economic growth. For example, the Bank of Russia took prompt action on any change in the real demand for money and took additional steps to stimulate positive economic dynamics, cut interest rates, damp down inflationary expectations, and slow the inflation rate, and as a result, the ruble gained somewhat in real terms and financial market stability increased.[26]

The Bank of Russia has given priority to some challenging problems for the bank. The press release states that one of the challenging problems is that specific risks connected with the dynamics of the prices of some financial assets and the prices situation on the real estate market have begun to play an increasingly important role. The practice of lending to related parties led to high- risk concentrations in some of the banks, which resulted in the Bank of Russia providing advancement of the methods of banking regulation and increasing management decision making and operation efficiency. The Bank of Russia was confronted with another problem identified as the fictitious capitalization of banks. In its decision to prevent banks from using all sorts of schemes designed to artificially overvalue or undervalue the required ratios, the Bank of Russia, in 2004, issued a number of regulations, especially the regulation that included the following: "On the Procedure for Creating Loan Loss Reserves by Credit Institutions" and the instruction "On Banks' Required Ratios."[27]

O. Bank of India, ranked first among the nationalized banks as India's most trusted, Mumbai-400023[28] (E-mail: mumbaisouthzone,dpo@bankofindia.co.in).

"The Bank of India has joined the National Commodity Derivatives Exchange of India Ltd. (NCDEX) as a depository participant offering commodity DEMAT Account Facility through the NSDL as well as CDSL DPOs," according to the Bank of India website announcement.[29] The Bank of India Stock Exchange Branch is one of the clearing banks for settlement of trade in commodities of the National Commodity Derivatives Exchange of India Ltd. (NCDEX). In addition, the Bullion Exchange Branch is a clearing bank of the Multi-Commodity Exchange of India Ltd. (MCX).[30]

The National Commodity & Derivatives Exchange Limited (NCDEX includes in its own description the statement that it is a professionally managed online multi-commodity exchange, and among its shareholders are large national-level institutions, large public-sector banks and companies, and Goldman Sachs.[31] In highlighting the dealings in derivatives contracts, the NCDEX authorized that all derivatives contract specifications shall be announced by the Exchange in advance from time to time. In addition, dealings shall be permitted on the Exchange in derivatives contracts pertaining to commodities as provided in the regulations, bylaws, and rules of the Exchange in force from time to time and for such categories of trading members, trade types, market types, settlement periods, and such trading hours as the Exchange may specify from time to time. In a demonstration of its management authority, the NCDEX at its discretion at any time could suspend trading in particular derivatives contracts as it deems appropriate. The Exchange may also revoke the suspension of trading in any types of derivatives contracts and/or commodities at any time.

P. The People's Bank of China.[32]

"Decree No. 532 of the State Council of the People's Republic of China: The Regulations on Foreign Exchange System of the People's Republic of China, Reviewed and Passed in the 20th Executive Meeting of the State Council on August 1, 2008, are hereby released and shall come into force simultaneously," Wen Jiabao, premier of the People's Republic of China, August 5, 2008.[33] An excerpt from these Regulations on the Foreign Exchange System of the People's Republic of China, which were issued in the Decree No. 193 of the State Council on January 28, 1996, amended in accordance with the decision of the State Council to review the Regulations on the Foreign Exchange System of the People's Republic of China on January 14, 1997 and revised and passed in the 20th Executive Meeting of the State Council on August 1, 2008, are formulated with a view to improving foreign exchange administration, while maintaining an equilibrium in the balance of payments and promoting sound economic development. Under chapter 1"General Provisions of These Regulations," it was established that the Foreign Exchange Administration Department of the State Council and the local offices (hereinafter referred to collectively as "the exchange administration agencies" shall exercise foreign exchange administration in accordance with the laws and regulations and shall assume the responsibility for the implementation of these regulations. It is significant that the specific issuance and transactions of securities or derivatives in the People's Republic of China by foreign entities or foreign individuals shall be made in accordance with the regulations governing market entry, and be registered in accordance with the regulations made by the Foreign Exchange Administration Department of the State Council, which is reported in chapter 3 of these regulations.[34]

Q. The People's Bank of China (PBC), Home—About PBC, is legally confirmed as The People's Bank of China's central bank status by the Law of the People's Republic of China on the People's Bank of China adopted on March 18, 1995 by the 3rd Plenum of the 8th National People's Congress. The People's Bank of China, No. 32 Chengfang Street, Xi Cheng District, Beijing, China 100800.[35]

With the focus on the improvement of the socialist market economic system, the People's Bank of China, in its capacity as the central bank of the People's Republic of China, is essential and possesses a significant role in China's macroeconomic management, according to the statement in the press release, Home—About PBC, from the People's Bank of China.[36]

Among the major functions that originated from and have been provided by the People's Bank of China are the following: (a) issuing the renminbi and administering its circulation, (b) managing the state treasury s fiscal agent, (c) regulating financial markets, (d) making payment and settlement rules and ensuring normal operation of the payment and settlement system, and (e) participating in international financial activity at the capacity of the central bank.[37]

In its report, "Guidance on Internal Control of the Commercial Banks (Draft Note)," the People's Bank of China prompted the commercial banks to take precautions against financial risks and guarantee safe and steady operation of the commercial banks banking system.[38]

The People's Bank of China has enumerated its enacted supervision and guidance of internal controls that consisted of 10 chapters and 144 articles. At this point in time of specific relevancy are Article 48 and Article 68. The People's Bank of China, under the provision of Article 48, ensures that the commercial bank shall strengthen the examination and supervision on the use of loans to prevent that the borrower "squeeze borrowers" from obtaining credit funds through loans, bank acceptance of bills, discount of interbank lending, changing the use of loans, and even investing the fund into the stock market, high-risk real estate, financial derivatives financing ramifications, and other illegal activities.[39]

The People's Bank of China, as part of its examination and supervision of guidance on internal controls of commercial banks, directed in Article 68 that the commercial banks shall clearly know the character and be clear about the nature of the financial derivatives, ramifications, risk discovering, and the decreasing strategies hedging tactics. The bargainers and dealers are directed to report to the senior management on the exchange trade details as possible contingent assets, hidden risks, and decreasing strategic hedging tactics in the financial ramifications of derivatives.[40]

R. South African Reserve Bank, the Central Bank of the Republic of South Africa–About Us–South Africa Reserve Bank; Guidelines–South African Reserve Bank, Pretoria, South Africa.[41]

The South African Reserve Bank is the central bank of the Republic of South Africa. An announcement by the South African Reserve Bank reported that the bank is responsible, on behalf of the Minister of Finance, for the day-to-day administration of exchange controls in South Africa. According to the bank announcement, the Exchange Control Regulations are legal provisions that limit the extent to which South African residents and companies may transfer funds abroad.[42] The bank's Financial Surveillance Department is, accordingly, responsible for the day-to-day administration of exchange controls.

S. International Monetary Fund (IMF), Financial Derivatives[43] (http://www.imf.org/external/np/sta/fd/).

In addressing the topic "Financial Derivatives," the International Monetary Fund defined financial derivatives as financial instruments that are linked to a specific financial instrument or indicator or commodity, and through which specific financial risks can be traded in financial markets in their own right.[44] The IMF position is that transactions in financial derivatives should be treated as separate transactions rather than as integral parts of the value of underlying transactions to which they may be linked. The IMF expressed the view that the value of a financial derivative derives from the price of an underlying item such as an asset or index. Unlike the debt instruments, there is no principal amount that is advanced to be repaid and no investment income accrued.

The IMF statement continued with the declaration that financial derivatives enable parties to trade specific financial risks such as interest rate risk, currency, equity, and commodity price risks, credit risk, etc. There are other entities, the IMF noted, that are more willing, or better suited, to take or manage these types of risks.[45]

The controversy and the intrigue concerning the definition and concepts of global over-the-counter (OTC) derivatives have been deepened based on the 1997 IMF discussion paper entitled "The Statistical Measurement of Financial Derivatives," which was adopted by the IMF Committee on Balance of

Payments Statistics and the Inter-Secretariat Working Group on National Accounts. This IMF discussion paper stressed the importance and emphasized the nature of financial derivatives; at the same time the discussion paper proposed that over-the-counter financial derivatives be treated as financial assets, and as a result, a change be made to the treatment of interest rate swaps and forward rate agreements (FRAs) so that instead of being recorded as financial assets, they be recorded in the financial account (and the outstanding positions be recorded in the international investment position).[46]

T. U.S. General Accounting Office (GAO), Report to Congressional Requesters, *Long-Term Capital Management: Regulators Need to Focus Greater Attention On Systemic Risk* (GAO/GGD-00-3), Washington, DC 20548, October 29, 1999.[47]

This report on Long-Term Capital Management's (LTCM) near collapse, in addition to the broader issues it has raised, is a response to a request from Congress. Between January and September 1998, according to the U.S. General Accounting Office, Long-Term Capital Management was one of the largest U.S. hedge funds, and it lost almost 90 percent of its capital.

The GAO Report established that in September 1998, the Federal Reserve determined that the rapid liquidation of LTCM's trading position and related positions of other market participants might pose a significant threat to the already unsettled global financial markets. Therefore, the Federal Reserve facilitated a private sector recapitalization to prevent LTCM's collapse.[48]

The Bank for International Settlements (BIS), according to the GAO Report, considered LTCM to be a "market maker" in some markets. For instance, some of its contemporaries' descriptions were that LTCM's counterparties often treated LTCM more like an investment bank instead of a hedge fund. There are some estimates provided by hedge fund researchers that suggested 70 percent of hedge funds use leverage, including a leverage ratio of less than two to one; however, leverage was an important part of LTCM's investment strategy.[49]

Hedge funds possess undercurrents of controversy. Even though there is no statutory definition of hedge funds, the GAO Report declared that the term "hedge funds" is commonly used to describe private investment vehicles that often engage in active trading of various types of securities and commodities. Although some funds are subject to certain federal reporting requirements, hedge funds are generally exempt from direct federal regulation. Derivatives are financial products, the GAO Report continued, whose value is determined from any underlying reference rate (interest rates, foreign currency exchange rates); index (reflects the collective value of various financial products); or asset (stocks, bonds, and commodities). In addition, derivatives can be (1) traded in central locations that are called exchanges, where buyers and sellers, or their representatives, meet to determine prices; and (2) privately negotiated by the parties that are off the exchange or over-the-counter (OTC). Leverage is broadly defined in the GAO Report as the ratio between some measure of risk and capital. The simple balance sheet leverage is assets divided by capital. However, this measure of leverage does not take into account the risk from off-balance sheet activities, such as the use of derivatives.[50]

U. U.S. Congress, House of Representatives, 111th Congress, 2nd Session, Report 111-517, *Dodd-Frank Wall Street Reform and Consumer Protection Act, Conference Report to Accompany H.R. 4173*, Washington, DC: U.S. Government Printing Office, June 29, 2010.[51]

The Dodd-Frank Act stated clearly that in reference to the "Ban on Proprietary Trading in Derivatives," an insured depositary institution shall comply with the prohibition on proprietary trading in derivatives as required in Section 619 of the Dodd-Frank Wall Street Reform and Consumer Protection Act.[52] An excerpt from Section 619 of Dodd-Frank is provided herewith:[53] "Sec, 619. Prohibition on Proprietary Trading and Certain Relationships with Hedge Funds. The Bank Holding Company Act of 1956 (12 U.S.C.1841 et seq.) is amended by adding the following: Sec.13. Prohibition on Proprietary Trading and Certain Relationships with Hedge Funds and Private Equity Funds:

(a) In General.—
  (1) Prohibition.—Unless otherwise provided in this section, a banking entity shall not
    (A) Engage in proprietary trading; or
    (B) Acquire or retain any equity, partnership, or other ownership interest in or sponsor a hedge fund or a private equity fund.
  (2) Nonbank Financial Companies Supervised by the Board.—Any nonbank financial company supervised by the Board that engages in proprietary trading or takes or retains any equity, partnership, or other ownership interest in or sponsors a hedge fund or a private equity fund shall be subject, by rule as provided in subsection (b)(2), to additional capital requirements for and additional quantitative limits with regards to such proprietary trading and taking or retaining any equity, partnership, or other ownership interest in or sponsorship of a hedge fund or a private equity fund, except that permitted activities as described in subsection (d) shall not be subject to the additional capital and additional quantitative limits except as provided in subsection (d)(3), as if the nonbank financial company supervised by the Board were a banking entity.

(b) Study and Rulemaking.—

(1) Study—Not later than 6 months after the date of enactment of this section, the Financial Stability Oversight Council shall study and make recommendations on implementing the provisions of this section . . ." The Dodd-Frank Act, in fact, demonstrates, describes, and presents a comprehensive statement on the swap scale and scope; for instance, the Dodd-Frank Act addresses the terminology in this manner:

"(47) SWAP—

(A) In General.—Except as provided in subparagraph (B), the term Swap means any agreement, contract, or transaction—

(i) That is a put, call, cap, floor, collar, or similar option of any kind that is for the purchase or sale, or based on the value, of one or more interest or other rates, currencies, commodities, securities, instruments of indebtedness, indices (indexes), quantitative measures, or other financial or economic interests or property of any kind;

(ii) that provides for any purchase, sale, payment, or delivery (other than a dividend on an equity security) that is dependent on the occurrence of an event or contingency associated with a potential financial, economic, or commercial consequence;

(iii) that provides on the executor basis for the exchange, on a fixed or contingent basis, of or more payments based on the value of one or more interest or other rates, currencies, commodities, securities, instruments of indebtedness, indices (indexes), quantitative measures, or other financial or economic interest therein or based on the value thereof, and that transfers, as between the parties to the transaction, in whole or in part, the financial risk associated with a future change in any such value or level without also conveying a current or future direct or indirect ownership interest in an asset (including any enterprise or investment pool) or liability that incorporates the financial risk so transferred, including any agreement, contract, or transaction commonly known as—

(I) an interest rate swap

(II) a rate floor

(III) a rate cap

(IV) a rate collar

(V) a cross-currency rate swap

(VI) a basis swap

(VII) a currency swap

(VIII) a foreign exchange swap
(IX) a total return swap
(X) an equity index swap
(XI) a total return swap
(XII) a debt index swap
(XIII) a debt swap
(XIV) a credit spread
(XV) a credit default swap
(XVI) a credit swap
(XVII) a weather swap
(XVIII) an energy swap
(XIX) a metal swap
(XX) an agricultural swap
(XXI) an emission swap; and
(XXII) a commodity swap

(iv) what is an agreement, contract, or transaction that is or in the future becomes, commonly known to the trade as a swap;
(v) Including any security-based swap agreement which meets the definition of 'swap agreement' as defined in Section 206A of the Gramm-Leach-Bliley Act (15 U.S.C. 78c note) of which a material term is based on the price, yield, value, or volatility of any security or any group or index of securities, or any interest therein; or
(vi) that is any combination or permutation of, or option on, any agreement, contract, or transaction described in any of clauses (i) through (v)."[54]

V. U.S. General Accounting Office (GAO), *Report to Congressional Committees Financial Derivatives Actions Taken or Proposed Since May 1994* (GAO/GGD/AIMD-97-8), Washington, DC, November 1996.[55]

It was in May 1994 that the GAO released its Report on Financial Derivatives that provided a description and risks. GAO made it known that derivatives are financial products whose value is determined from an underlying reference rate, index, or asset. The composition of the underlying consists of stocks, bonds, commodities, interest rates, foreign currency exchange rates, and indexes that reflect the collective value of various financial products.[56]

W. Derivatives Instruments Sample Bank Holding Company FRB District Peer: 1, *BHCPR User's Guide,* March 2011[57] (http://www.frb.gov).

The narrative statement on derivatives instruments on page 10 of the *BHCPR User's Guide* provides the dollar amounts of derivative contracts, which are grouped according to underlying risk exposure (interest rate contracts, foreign exchange contracts, equity derivative contracts, and commodity and other derivative contracts), financial technique used (futures, options, or swaps), and purpose (trading or hedging). Additional derivatives information includes the gross positive or negative fair value, current credit exposure of derivatives covered under the risk-based capital standards, maturity distribution, replacement costs of past-due derivative instruments, and impact of non-traded derivative contracts on net income.[58]

NOTES

1. U.S. Commodity Futures Trading Commission (CFTC), CFTC Glossary (http://www.cftc.gov/consumerprotection/educationcenter/cftcglossary/glossary_o).
2. *Over-the-Counter* [off the exchanges] *Derivatives Markets and the Commodity Exchange Act*, Report of the President's Working Group on Financial Markets, Department of the Treasury, Board of Governors of the Federal Reserve System, Securities and Exchange Commission, Commodity Futures Trading Commission, November 1999, p. 4.
3. U.S. Department of the Treasury Resource Center, U.S. Derivatives Contracts with Foreigners, Holdings and Transaction (http://www.treasury.gov/resource-center/data-chart-center/tic/pages/derivativesd.aspx4/23/2012).
4. Stephanie E. Curcuru (with assistance of Jonas J. Robison), "U.S. Cross-Border Derivatives Data: A User's Guide," Federal Reserve Bulletin Board's Division of International Finance, May 2007, p. A1.
5. Ibid., p. 2.
6. Gary Gensler, Keynote address of Chairman Gary Gensler, "OTC Derivatives Reform, Market's Outlook," OTC Derivatives Markets Conference, Washington, DC, March 9, 2010, p. 16.
7. Federal Reserve Bank of New York Staff Report, *An Analysis of OTC Interest Rate Derivatives Transactions: Implications for Pubic Reporting*, Staff Report No. 1557, by Michael Fleming, John Jackson, Ada Li, Asani Sarkar, and Patricia Zobel, New York: Federal Reserve Bank of New York, March 2012, pp. 2–4. Patrick M. Parkinson, Deputy Director, Division of Research and Statistics, "Over-the-Counter Derivatives," before the Subcommittee on Securities, Insurance, and Investment, Committee on Banking, Housing, and Urban Affairs, U.S. Senate, Washington, DC: Board of Governors of the Federal Reserve System, July 9, 2008, pp. 1–2 (http://www.federalreserve.gov/newseents/testimony/parkinson2008, 0709a.htm.4/3/2012).
8. Larry H. Summers, Secretary, U.S. Department of the Treasury; Alan Greenspan, Chairman, Board of Governors of the Federal Reserve System; Arthur Levitt, Chairman, U.S. Securities and Exchange Commission; William J. Rainer, Chairman, U.S. Commodity Futures Trading Commission, *Over-the-Counter Derivatives*

Markets and the Commodity Exchange Act, Report of the President's Working Group on Financial Markets, November 1999, passim.

9. Testimony of Kevin Warsh, Board of Governors of the Federal Reserve System, "Hedge Funds," before the Committee on Financial Services, U.S. House of Representatives, Washington, DC, July 11, 2007, passim.

10. Carl Levin, Chairman, and Tom Coburn, Ranking Minority Member, U.S. Senate, Permanent Subcommittee on Investigations, Committee on Homeland Security and Governmental Affairs, *Wall Street and the Financial Crisis: Anatomy of a Financial Collapse*, Majority and Minority Staff Report, April 13, 2011, p. 2.

11. Ibid., pp. 2–4.

12. Ibid., pp. 9–10.

13. Ibid., p. 386.

14. Ibid., pp. 386–387.

15. Central Bank of Japan, "The Bank of Japan's Initiatives on the Macroprudential Front, October 18, 2011."

16. Ibid., p. 1.

17. Ibid., p. 2.

18. Ibid., pp. 7–8.

19. U.S. Securities and Exchange Commission, "SEC Adopts Rule Defining Swaps-Related Terms for Regulating Derivatives," April 18, 2012 (http://www.sec.gov/news/press/2012//2012-67.htm).

20. Ibid., pp. 1–2.

21. Ben Protess, "Regulators to Ease a Rule on Derivatives Dealers," *New York Times*, April 18, 2012, p. B1.

22. The Banco Central do Brasil (Central Bank of Brazil), a federal agency member of the National Financial System (http://www.bcb.gov/?bs 200212en).

23. http://www.bcb.gov.br/?derivatives(bm&fbovespaderivatives).

24. Bank of Russia Today, Bank of Russia, the Central Bank of the Russian Federation, Moscow, Russia, (webmaster@www.cbr.ruhttp://www.cbr.ru/eng/today/default.aspx).

25. Ibid., p. 2.

26. Ibid., p. 3.

27. Bank of India, Depository Services, cdsl/nsdl charge; nsdl clearing member charges; cdsl clearing member charges, Mumbai–400023 (http://www.bankofindia.com/depositservice3.aspx).

28. Ibid., p. 1.

29. Ibid.

30. National Commodity & Derivatives Exchange Limited (NCDEX), (http://www.ncdex.com/aboutus/profile.aspx).

31. 1."Title, Extent and Commencement, The Regulations Framed Hereunder Shall be Known as National Commodity & Derivatives Exchange Limited, Regulations, 2003 (herein referred to as 'Regulations') and shall come into effect immediately on approval by the Forward Market Commission or any other authority appointed under the Forward Contracts (Regulations) Act, 1952 or any other applicable law"; 5.1 Dealing in Derivatives Contracts, p. 27.

32. The People's Bank of China, regulations on foreign exchange system of the People's Republic of China, Decree No. 532 of the State Council of the People's Republic of China; the regulations on foreign exchange system of the People's Republic of China, reviewed and passed in the 20th Executive Meeting of the State Council on August 1, 2008, are hereby released and shall come into force simultaneously. Wen Jiabao, premier of the People's Republic of China, August 5, 2008 (http://www.pbc.gov.cn/publish/english/964/1958/1958/_html).

33. Ibid., chapter 3, p. 4.

34. The People's Bank of China, E.mail: webbox@pbc.gov.cn Address: No. 32 Chengfang Street Xi Cheng District, Beijing, China 100800 (http://www.pbc.gov.cn/publish/english/952/index.html).

35. Ibid.

36. Ibid. (See also the People's Bank of China, Guidance on Internal Control of the Commercial Banks (draft note). (http://www.pbc.gov.cn/publish/english/964/1956/1963/19563_html).

37. The People's Bank of China, Guidance on Internal Control of the Commercial Banks (draft note), No. 32 Chengfang Street, Xi Cheng District, Beijing, China 100800. E.mail: webbox@pbc.gov.cn, (http://www.pbc.gov.cn?publish/english/1952/index.html).

38. Ibid., p. 7.

39. Ibid., p. 10.

40. South African Reserve Bank, the Central Bank of the Republic of South Africa, "Financial Surveillance and Exchange Controls," Praetoria, South Africa (http://www.resbabk.co.za/aboutus/pages/about%/20us_home.aspx).

41. South African Reserve Bank, Financial Surveillance and Exchange Controls (http://www.resbank.co.za/regultionandsupervision/financialsurveillanceandexchangeC…).

42. International Monetary Fund (IMF), Financial Derivatives, IMF Committee on Balance of Payment Statistics: Financial Derivatives (http://www.imf.org/external/np/sta/fd/).

43. Ibid., p. 1.

44. Ibid.

45. Ibid., pp. 1–2.

46. U.S. General Accounting Office (GAO), *Report to Congressional Requesters, Long-Term Capital Management: Regulators Need to Focus Greater Attention on Systemic Risk* (GAO/GGD-00-3), Washington, DC 29548, October 29, 1999.

47. Ibid., p. 1.

48. Ibid., p. 5.

49. Ibid., pp. 1 and 6 (See also Appendix 1, pp. 44–45).

50. U.S. Congress, House of Representatives, 111th Congress, 2nd Session, Report 111-517, *Dodd-Frank Wall Street Reform and Consumer Protection Act, Conference Report to Accompany HR 4173*, Washington, DC: U.S. Government Printing Office, June 29, 2010.

51. Ibid., p. 281.

52. Ibid., p. 250.

53. Ibid., pp. 296–297.

54. U.S. General Accounting Office (GAO) Report to Congressional Committees, *Financial Derivatives Actions Taken or Proposed Since May 1994* (GAO/GGD/AIMD-97-8, Washington, DC, November 1996.

55. Ibid., p. 3.

56. Sample Derivative Instrument, *BHCPR User's Guide*, March 2011, sec. 3-43 (FRB.gov).

57. Ibid., sec. 3-43 to 3-46.

58. House of Representatives, 106th Congress, 1st Session, Gramm-Leach-Bliley Act, Report 106-434, November 2, 1999, Sec. 101. Glass-Steagall Act Repeal, p. 5.

3

Grand Tying Arrangements and Reciprocal Dealings (Reciprocity)

The Gramm-Leach-Bliley Act (GLB Act), enacted n 1999, repealed Section 20 of the Banking Act of 1933 (12 U.S.C. 377) and repealed Section 32 of the Banking Act of 1933 (12 U.S.C. 78)[1], commonly referred to as the "Glass-Steagall Act" which was enacted by Congress in 1933 and instituted the direct and specific separation between commercial banking and affiliation, and investment banking. In effect, Glass-Steagall directed the commercial banks' termination of their securities affiliates, Section 20, and management interlock, Section 32.

Under the provisions of the Gramm-Leach-Bliley Act (1999), there is heightened emphasis placed on allowing banks, securities broker-dealers, and insurance companies to affiliate with each other through the bank holding company structure.[2]

Chapter 3 is the author's assessment of the grand alliance based on the institutional arrangements, organizational characteristics, and business conduct performance under provisions of the joint Gramm-Leach-Bliley Act (1999) and Board of Governors of the Federal Reserve System strategic agglomeration movement.

The GLB Act structure and the significance of its powerful influence relate directly to Title I— Facilitating Affiliation Among Banks, Securities Firms, and Insurance Companies; Subtitle A—Affiliations. For example, there are two critical issues that make a contribution to the consolidation actions and are elevated to priority status—namely, (1) Section 102: Activity Restrictions Available to Bank Holding Companies that Are Not Financial Holding Companies; and (2) Section 103: Financial Activities—Engaging in Activities That Are Financial in Nature.

Section 102 of the GLB Act (1999) is an amendment to Section 4(c)(8) of the Bank Holding Company Act of 1956 (12 U.S.C. 1843)(c)(8) by adding the following statement:

(a) IN GENERAL—

Section 4(c)(8) of the Bank Holding Company Act of 1956 (12 U.S.C. 1843(c)(8) is amended to read as follows:

"(8) shares of any company the activities of which had been determined by the Board by regulation or order under this paragraph as of the day before the date of the enactment of the Gramm-Leach-Bliley Act, to be so closely related to banking as to be a proper incident thereto (subject to such terms and conditions contained in such regulation or order, unless modified by the Board)."

(b) CONFORMING CHANGES TO OTHER STATUTES.—

(1) AMENDMENT TO THE BANK HOLDING COMPANY ACT

AMENDMENTS OF 1970—Section 105 of the Bank Holding Company Act Amendments of 1970 (12U.S.C.1850) is amended by striking, "to engage directly or indirectly in a nonbanking activity pursuant to Section 4 of such Act"

(1) AMENDMENT TO THE BANK SERVICE COMPANY.ACT—Section 4(f) of the Bank Service Company Act (12U.S.C.1864(f) is amended by inserting before the period at the end the following: "as of the day before the date of the enactment of the Gramm-Leach-Bliley Act."

Section 103, Financial Activities, Title I of the Gramm-Leach-Bliley Act (1999) is the creation of the financial holding company source for the determination and dominance of the monetary, financial, and Wall Street market structure organizational characteristics.

In the author's assessment, the essential ingredients and salient features that result from Section 103, Financial Activities applications, in fact, place substantial emphasis on (a) a financial holding company determination of the degree of seller concentration in the market,(b) a financial holding company definition of the number and size distribution of buyers in the market, (c) a financial holding company determination of the degree of global over-the-counter (OTC) derivatives instruments output differentiation among the various sellers in the market, (d) a financial holding company determination of condition of freedom of entry in the global over-the-counter derivatives instruments market is interdependent with the established financial holding company advantages accrued over the potential entrants, and (e) a financial holding company determination of vertical integration, horizontal integration, interlocking directorates, and conglomeration of industries, in effect, fosters concentration and control of global over-the-counter derivatives market structure and conduct.

The Gramm-Leach-Bliley Act (1999), Title I, Section 103, Financial Activities added new subsections in the amendment to "IN GENERAL.—Section 4 of the Bank Holding Company Act of 1956 (12 U.S.C. 1843): Engaging In Activities That Are Financial In Nature." For instance, a financial holding company may engage in any activity that the Federal Reserve Board (FRB), in accordance with paragraph (2), determines (by regulation or order) to be financial in nature or incidental to such financial activity, and is complementary to a financial activity and does not pose a substantial risk to the safety or soundness of depository institutions or the financial system generally.

Numerous factors are considered in engaging in activities that are financial in nature, i.e., in the determination of whether an activity is financial in nature or incidental to a financial activity, the FRB shall take into account such issues as the purpose of this Act (the Bank Holding Company Act of 1956 [12 U.S.C. 1843]) and the Gramm-Leach-Bliley Act (1999); take into account changes or reasonably expected changes in the marketplace in which financial holding companies compete; and take into account changes or reasonably expected changes in the technology for the delivery of financial services.

In determining activities that are financial in nature, the following activities shall be considered to be financial in nature: lending, exchanging, transferring, investing for others, or safeguarding money or securities; insuring, guaranteeing, or indemnifying against loss, harm, damage, illness, disability, or death, or providing and issuing annuities, and acting as principal, agent, or broker for purposes of the foregoing, in any state; underwriting, dealing in, or making a market in securities; providing financial, investment, or economic advisory services, including advising on an investment company (as defined in Section 3 of the Investment Company Act of 1940); engaging in any activity that the FRB has determined, by order of regulation that is in effect on the date of the enactment of the Gramm-Leach-Bliley Act (1999), to be so closely related to banking or managing or controlling banks as to be a proper incident thereto; and engaging, in the United States, in any activity that (1) a bank holding company may engage in outside of the United States, and (2) the FRB has determined, under regulations prescribed or interpretations issued pursuant to Subsection (c)(13) (as in effect on the day before the date of the enactment of the Gramm-Leach-Bliley Act), to be usual in connection with the transaction of banking or other financial operations abroad. The financial holding company, when examined at its core, places emphasis on the dominance of centralization, consolidation, concentration of control, interlocking directorates, financial ties, and money and credit. This is the author's assessment.

ROLE OF THE BOARD OF GOVERNORS OF THE FEDERAL RESERVE SYSTEM, FINANCIAL HOLDING COMPANIES, AND THE GRAMM-LEACH-BLILEY ACT

The Board of Governors of the Federal Reserve System released the decision on financial holding companies as of May 18, 2012. The Board announced that under the Bank Holding Company Act provisions, bank holding companies may elect to be financial holding companies. The Board's determination centered on its list of those companies, in fact, the 488 financial holding companies whose elections to become or be treated as financial holding companies were effective as of May 18, 2012;[3] one specification clearly established that the list includes only the top-tier company in each of the 488 listed financial holding companies.

This formation of aggregate concentration of control of financial holding companies approved by the Board of Governors of the Federal Reserve System, in effect, facilitated the centralization of the determinants of market structure; enlarged the number and size distribution of financial holding companies; created variations and huge formations of global over-the-counter (OTC) derivatives financial holdings; increased barriers to entry in the financial holding companies line of business; and consolidated the high degree of vertical integration, horizontal integration and conglomerate size.

The bank holding companies also may own broker-dealers engaged in securities activities or insurance companies, according to the Federal Reserve "Purposes and Functions and Umbrella Supervision and Coordination with Other Functional Regulators."[4]

One of the fundamental provisions of the Gramm-Leach-Bliley Act (1999) is the authorization of banks, securities broker-dealers, and insurance companies to interconnect activities, affiliation, and operations through the bank holding company structure, in which a bank holding company must meet certain capital, management, and other requirements and, specifically, must elect to become a "Financial Holding Company." Then, the Federal Reserve's supervisory responsibility and authority, particularly as related to supervision and coordination with other functional regulators, includes conducting performance supervision and appraisal of a bank holding company or financial holding company as well as reviewing and assessing the consolidated organization's operations, risk-management systems, and capital adequacy.[5]

It is essential to obtain first the approval from the Federal Reserve if a firm should seek to become a bank holding company under the Bank Holding Company Act and its definition of bank holding company as any company that directly or indirectly owns, controls, or has the power to vote 25 percent or more of any class of the voting shares of a bank; controls in any manner the election of a majority of the directors or trustees of a bank; or is found to exercise a controlling influence over the management or policies of a bank.

A bank holding company must obtain the approval of the Federal Reserve before acquiring more than 5 percent of the shares of an additional bank holding company.[6]

The grand tying arrangement and reciprocity, in essence, have been driven by and embodied in the Gramm-Leach-Bliley Act (the seller of global over-the-counter [OTC] derivatives contracts/credit default swaps), which repealed the depression-era Glass-Steagall Act enacted by Congress in 1933 that ordered and forced the separation between commercial banking and investment banking. In contrast and under conditions of financial modernization and restructuring sub rosa, the Gramm-Leach-Bliley Act was enacted by Congress in 1999 and permitted banks to affiliate with securities firms and insurance companies within financial holding companies; allowed a bank holding company that qualifies to become a financial holding company to, in fact, engage in a broader range of financially related activities, among which are full-scope securities underwriting and dealing, insurance underwriting and sales, and merchant banking; required a bank holding company or financial holding company to obtain the approval of the Federal Reserve before acquiring more than 5 percent of the shares of an additional bank or bank holding company.

This type of tying contract and form of reciprocity, for instance, are embedded in the tie-ins products reported and authorized by the Board of Governors of the Federal Reserve System's list of the 488 financial holding companies whose election to become or be treated as financial holding companies is effective as of May 18, 2012.[7]

The grand tying arrangement and reciprocity enabled under the Gramm-Leach-Bliley Act ("Producer") and the Board of Governors of the Federal Reserve System ("Buyer") possess potential prohibition of other sellers of the tied-in product, which is global OTC derivatives contracts/credit default swaps ("Contracts"), from engaging in required freedom of entry. Thus, prohibition has the effect of altering the impact on the market for the tied-in product and contributing to the condition that would tend to substantially lessen competition or create monopoly, activate exclusive dealing, or constitute an unreasonable restraint of trade under the Sherman Act, Section 1 and Section 2.

MODEL OF GRAND TYING ARRANGEMENT AND RECIPROCAL DEALING (RECIPROCITY)

A standard characteristic and frequent feature of most tying arrangements or contracts hold that there are some restrictive practices on competition that cover various (and usually adjacent) vertical stages of an agreement between manufacturers and their distributors, or, namely, vertical restrictions. What is inherent in tying arrangements or contracts, moreover, is that the purchaser of some article—e.g., over-the-counter derivatives—agrees to a condition of purchase to buy the seller's supplies of some other commodity, such as a credit default swap. As a result, the agreement forecloses sellers' competition and blocks the opportunity of selling the tied financial product to that purchaser.

Exclusive dealing is a critical component of tying arrangements or contracts under which some company agrees to sell only the financial products of one particular firm and, thereby, agrees implicitly or explicitly not to handle the financial products of competing companies; thus, the suppression of competition.

There are some companies that possess a wide array of reasons for attempting to tie the sale of one financial product to that of another. One reason is that tying is frequently a convenient way of price-fixing as a barrier to the freedom of entry. The lodestone of grand tying arrangements, for example, has been driven by the formation and implementation of the Grand Alliance of DuPont and Imperial Chemical Industries (ICI). The 1939 DuPont-ICI Patents and Processes Agreement, dated December 1938, according to *U.S. v. Imperial Chemical Industries, Ltd., et al* in U.S. District Court for the Southern District of New York, Civil Action No. 24-13, Exhibit 2 in Complaint Filed January 6, 1944. This 1939 General Alliance provided for technical and secret information. When the previous 1929 Patents and Processes Agreement, as amended in 1934, expired in 1939, the Grand Alliance of Du Pont and Imperial Chemical Industries was successfully renewed in 1939 but covered only 50 to 60 percent of their production. In addition, the 1929 Patents and Processes Agreement provided for the exchange of technical information and patents—dividing markets by assigning to each member the patent rights for its allotted territory.

The lodestone of grand tying arrangements is emboldened by the *European Parliament Report on the Existence of a Global System for the Interception of Private and Commercial Communications (ECHELON Interception System), (2001/2098(INI))*, FINAL A5-0264/2001, PAR 1, Temporary Committee on the ECHLON Interception System, RR\445698EN.doc. There is a "marriage" between global over-the-counter (OTC) derivatives accumulation and the ECHELON, which is an interception system that differs from other intelligence systems in that it possesses two features which make it quite unusual. First is its capacity to carry out quasi-total surveillance. For example, satellite receiver stations and spy satellites in particular are alleged to give it the ability to intercept any telephone, fax, internet or e-mail message sent by any individual and thus to inspect its contents. Second, the unusual feature of ECHELON is said to be that the system operates worldwide, on the basis of cooperation proportionate to their capabilities, among several states (the UK, the United States. Canada, Australia, and New Zealand), giving it an added value in comparison to national systems: the states participating in ECHELON (the UK, U.S. states) can place their interception systems at each other's disposal, share the cost, and make joint use of the resulting information. This type of international cooperation is essential in particular for the worldwide interception of satellite communications, since only in this way is it possible to ensure in international communications that both sides of a dialogue can be intercepted. It is clear that, in view of its size, a satellite receiver station cannot be established on the territory of a state without that state's knowledge. Mutual agreement and proportionate cooperation among several states in different parts of the world are essential.

In effect, a tying arrangement exists when an underwriter of a desired financial product sells it only to those who also buy a second financial product from it. Moreover, in the traditional terminology, the underwriter ties the sale of the tying product (the desired financial product) to sales of the tied product (second financial product). The tie-in could prohibit other sellers of the tied product from an opportunity to compete for customers on the independent merits of the tied product itself and without the intervention of the tying product.

The tying arrangement might often constitute conditions of monopolization under the Sherman Act, Section 2; alternatively, a tying arrangement in the form of an agreement might be an unreasonable restraint of trade under the Sherman Act, Section 1. When two financial products are tied and sold as a package, the seller's receipts and the buyer's payments, in essence, reflect the aggregate transaction but might not be apportionably allocated among the components; thus, a tie-in transaction could be used to evade price controls and conceal the actual price of using the tying or tied product.

The reciprocal dealing, or reciprocity, encompasses the practice through which a company buys only from those companies who buy from it or purchases more from those who purchase from it. For instance, a company could increase its tendency toward engaging in reciprocal dealing (reciprocity) through diversification by mergers, acquisitions, or internal expansion, especially into products that its suppliers buy. Similar to the grand tying arrangements, the reciprocal dealing (reciprocity) practices might permit more effective exploitation of existing concentration of control and interlocking directorates, and harmful effects involving future competitive behavior.

THE MODEL

(a) Producer (Reciprocity): Gramm-Leach-Bliley Act (1999) increased the range of affiliations permitted to banking and financial institutions. The GLB Act (1999) repealed Section 20 (Affiliation) of the Banking Act of 1933, commonly referred to as the "Glass-Steagall Act," and repealed Section 32 (Management Interlock) of the Banking Act of 1933 that forced a separation between commercial banking and investment banking. The GLB Act permitted banks, securities broker-dealers, and insurance companies to affiliate with each other within financial holding companies;

(b) Producer/Seller (Reciprocity): Board of Governors of the Federal Reserve System. Under the Bank Holding Company Act, the bank holding companies under the authority of the Fed may elect to be financial holding companies. A list of 488 of those financial holding companies whose elections to become or be treated as financial holding companies are effective as of May 18, 2012 by the Board of Governors of the Federal Reserve System (http://www.federalreserve.gov/bankinforeg/the.htm).

(c) Buyer (Reciprocity): Board of Governors of the Federal Reserve System. Specifically with regard to the Gramm-Leach-Bliley Act (1999), the Federal Reserve Board is the only banking agency that has the authority to exempt any bank from GLB Act requirements. The Federal Reserve's role as the supervisor of a bank holding company or financial holding company is to review and assess the consolidated organizations operations, risk management systems, and capital adequacy to ensure that the holding company and its nonbank subsidiaries do not threaten the viability of the company's depository institutions. The Federal Reserve is the "umbrella supervisor" of the consolidated organizations.

(d) Tie-in Products: In this connection, and by definition, tie-in products are as follows:

   (1) bank holding companies or financial holding companies who elect to be financial holding companies under the Bank Holding Company Act of 1956 as amended and made effective by the Board of Governors of the Federal Reserve System, and included in Regulatory Federal Reserve Districts;

(2) global over-the-counter (OTC) derivatives contracts/credit default swap agreements; and
(3) financial products whose value is determined from an underlying reference rate, indices (indexes), analysis variance, ratios, or assets in which the underlying include stocks, bonds, commodities, foreign currency exchange rates, and indexes that reflect the collective value of various financial products.

(e) Degree of Seller (one, few, or many) Concentration in Tied Products: For example,
(1) he number and size distribution of the largest four seller companies in tied products determination of competition;
(2) the number and size distribution of the largest eight seller companies in tied products determination of competition; and
(3) degree of tied product differentiation, nonidentical by buyers, as among the outputs of the various sellers in the market in the determination of competition (tied products of competing sellers are identical or whether buyers have numerous or various choices).

(f) Patterns of Sellers Conduct and Behavior: These elements mean intercompany seller behavior and coordination of tied-product price and inter-seller sales promotion policies effect on the limits of potential competition. In the adjustment to competition, however, does the pattern of sellers conduct and behavior uncover tacit collusion or overt agreements on setting prices, exclusive dealing practiced against the freedom of entry,; and demonstration of predatory or exclusionary tactics in restraining trade; in sum, the elimination of barrier to entry and potential competition are included in the meaning of patterns of sellers conduct and behavior.

Under the grand tying arrangements and reciprocal dealings (reciprocity), the Gramm-Leach-Bliley Act (1999), and the Board of Governors of the Federal Reserve System, in essence "The Grand Alliance," the purchaser of global over-the-counter (OTC) derivatives enters an agreement (contract) as a condition of purchase to buy the seller's insurance (credit default swap). The contract, in effect, prohibits competing sellers from the opportunity of selling the tied product to that buyer, thus foreclosing competition and establishing an unreasonable restraint of trade.

NOTES

1. *The Federal Reserve System: Purposes & Functions*, 9th Edition, Washington, DC: Board of Governors of the Federal Reserve System, 2005, p. 65.

2. Board of Governors of the Federal Reserve System, "Financial Holding Companies," Washington, DC, May 18, 2012, p. 116 (http://www.federalreserve.gov/bankinforeg/the.htm).

3. *The Federal Reserve System: Purposes & Functions*, 9th Edition, Washington, DC: Board of Governors of the Federal Reserve System, 2005, p. 65.

4. Ibid.

5. Ibid., pp. 71–72.

6. Board of Governors of the Federal Reserve System, "Financial Holding Companies," Washington, DC, May 18, 2012, passim (http://www.federalreserve.gov/bankinforeg/the.htm).

7. Table 4.1 a-b: Global OTC Derivatives Accumulation Positions Market is compiled from the Bank for International Settlements' (BIS) numerous published sources, among which are BIS press releases, "The Global OTC Derivatives Market at End-December 1999 (http://www.bis.org/press/p990602.htm); BIS Quarterly Review, June 2012, Table 19: 2009, 2010, and 2011; U.S. General Accounting Office (GAO) Report to Congressional Committees, "Financial Derivatives Actions Taken or Proposed Since May 1994," (GAO/GGD/AIMD-97-8), November 1996, p. 3 (for 1995); Bank for International Settlements Monetary and Economic Department, Triennial and Semiannual Surveys Positions in Global Over-the-Counter (OTC) Derivatives Markets at end-June 2010, Basil, Switzerland, November 2010; and Bank for International Settlements, Triennial Central Bank Survey, Foreign Exchange and Derivatives Market Activity in 2007, Basil, Switzerland, December 2007.

4

Global Over-the-Counter (OTC) Derivatives Accumulation

In 1994 the U.S. General Accounting Office (GAO) issued its publication, *Financial Derivatives: Actions Needed to Protect the Financial System* (GAO/GGD-94-133, May 18, 1994), which emphasized the importance of financial derivatives that are characterized by numerous risks connected with four basic types of derivative products: (1) forwards, (2) futures, (3) options, and (4) swaps.

In addition, in 1996 the GAO published its report, *Financial Derivatives Action Taken or Proposed Since May 1994* (GAO/GGD/AIMD-97-8), which is a follow-up review of the assessment of the improvements, if any, that have been made by financial regulators and business professionals in the appraisal of the problem areas previously specified by GAO in its May 1994 report.

The GAO issued its publication, *Long-Term Capital Management: Regulators Need to Focus Greater Attention on Systemic Risk* (GAO/GGD-00-3, October 1999), which focused on questions such as the following: (a) how Long-Term Capital Management's (LTCM) positions became large and leveraged enough to be deemed a potential systemic threat, (b) what federal regulators knew about LTCM and when they found out about its problems, (c) what the extent of coordination among regulators was, and (d) whether regulatory authority limits regulators' ability to identify and mitigate potential systemic risk.

In November 1999 the Gramm-Leach-Bliley Act was enacted by Congress and under Section 101, Glass-Steagall Act, in fact, repeals (a) Section 20 Repealed—Section 20 of the Banking Act of 1933 (12 U.S.C. 377), commonly referred to as the "Glass-Steagall Act," is repealed; (b) Section 32 Repealed—Section 32 of the Banking Act of 1933 (12 U.S.C. 78) is repealed.

The underlying financial, banking, economic operations, trading transactions, and capital flows, especially huge global over-the-counter (OTC) derivatives accumulation, are driven by processes of concentration of control, market power, major voting rights in corporations, and interlocking directorates among dominant corporations.

The global over-the-counter (OTC) derivatives accumulation, which registered at $55.7 trillion on March 31, 1955, soared to the huge amount of $683.7 trillion at the end of June 2008, rising to the new gigantic level of $706.8 trillion at the end of June 2011, as released and presented in the following Table 4.1 a-b: Global Over-the-Counter (OTC) Derivatives Accumulation Position Market; compilation from numerous sources published by the Bank for International Settlements.

The Bank for International Settlements (BIS), which was estabished on May 17, 1930, serves as a bank for central banks and its capital is held by central banks only. BIS offers a wide range of financial services specifically designed to assist central banks and other monetary authorities in the management of their foreign exchange reserves.

Table 4.1a. Global Over-the-Counter (OTC) Derivatives Accumulation Position Market[1]

Year	*Position at End-June*	*Position at End-December*
1995	$55.7 Trillion (Mar. 3, 1995)	
1996		
1997		
1998	$72 Trillion	$80.3 Trillion
1999	$81.5 Trillion	$88.2 Trillion
2000	$94.0 Trillion	$95.2 Trillion
2001	$100.0 Trillion	$111.0 Trillion
2002	$128.0 Trillion	$141.7 Trillion
2003	$169.7 Trillion	$200.0 Trillion
2004	$220.1 Trillion	$248.0 Trillion
2005	$270.0 Trillion	$285.0 Trillion

Table 4.1b. Global Over-the-Counter (OTC) Derivatives Accumulation Position Market

Year	*Position at End-June*	*Position at End-December*
2006	$370.0 Trillion	$415.0 Trillion
2007	$507.9 Trillion	$507.9 Trillion
2008	$683.7 Trillion	$592.0 Trillion
2009	$605.0 Trillion	$603.9 Trillion
2010	$582.7 Trillion	$601.0 Trillion
2011	$706.8 Trillion	$647.7 Trillion

Thoughout the book the significance of Table 4.1a-b: Global Over-the-Counter (OTC) Derivatives Accummulation Position Market, and its application will serve as a lodestone in the assessment of the impact of huge global OTC derivatives accumulation on currencies, banking, financial, industrial, and commercial catastrophic losses and economic collapse.

During the Bank for International Settlements (BIS) implementation and documentation of the gigantic volume and huge notional amounts of global over-the-counter (OTC) derivatives trading, history, and operations, moreover, the U.S. General Accounting Office (GAO), in recognition of the galvanization of global OTC derivatives, issued three important reports: (1) *Financial Derivatives: Actions Taken or Proposed Since May 1994*, GAO/GGD/AIMD-97-8, November 1996; (2) *Long-Term Capital Management: Regulators Need to Focus Greater Attention on Systemic Risk*, GAO/GGD-00-3, October 1998; and (3) *Financial Derivatives: Actions Needed to Protect the Financial System*, GAO/GGD-94-133, May 18, 1994.

The accumulation of the huge volume and amount of global OTC derivatives detailed in the preceding Table 4.1 a-b are, in effect, driven by, accompanied by, and embedded in processes of concentration of control of global OTC derivatives through ownership, consolidations and mergers of banks, derivatives broker-dealers and centralization of management; major voting stock acquisitions in large U.S. corporations and competitors by the same centralization of management methods of acquiring control; increased concentration of control of global OTC derivatives through interlocking directorates and through joint ventures, consortia, partnerships, and holding companies engaged in multinational enterprise diplomacy.

This system of processes of concentration and control unfolded and began with the current assessment of the previously addressed publication, namely, the GAO Report to Congressional Committees, *Financial Derivatives: Actions Taken or Proposed Since May 1994*, GAO/GGD/AIMD-99-8.

The rapid expansion and acceleration in the accumulation of OTC derivatives and the galvanization formed by core institutions represent the significance of the framework presentation provided in the GAO's *Financial*

Table 4.2. Fifteen (15) Major U. S. Over-the-Counter (OTC) Derivatives Dealers

J. P. Morgan & Co., Inc.	Salomon, Inc.
The Chase Manhattan Corporation	Merrill Lynch & Co., Inc.
Citicorp	Morgan Stanley Group, Inc.
Bankers Trust New York Corporation	Lehman Brothers
BankAmerica Corporation	American International Group, Inc.
NationsBank Corporation	The Prudential Insurance Company
First Chicago Corporation of America	General Re Corporation
The Goldman Sachs Group, L.P.	

Table 4.3. Notional/Contract Amounts of Derivatives Reported by the 15 Major U. S. OTC Derivatives Dealers Year-End 1990 Through Year-End 1995[4] Dollars in Billions

Dealers	*1990*	*1991*	*1992*	*1993*	*1994*	*1995*
Banks (7)	$5,350	$5,811	$7,574	$10,353	$13,724	$15,809
Securities Firms (5)	1,730	2,188	2,967	4,474	5,880	6,966
Insurance Companies (3)	193	257	403	634	798	985
Total (15)	$7,273	$8,256	$10,944	$15,461	$20,402	$23,760

Derivatives Action Taken or Proposed Since May 1994. For instance, there were fifteen major U.S. OTC derivatives dealers year-end 1990 through year-end 1995 measured by notional/contract amounts of derivatives reported.[3] The fifteen major U.S. OTC derivatives dealers and contributors in the processes of concentration of control are enumerated in table 4.2.

In every year based on Table 4.3: Table 1.1 Notional/Contract Amounts of Derivatives Reported by the 15 Major U.S. OTC Derivatives Dealers Year-End 1990 Through Year-End 1995, the total derivatives volume of trading transactions increased from $7,273 billion in 1990 to $23,760 billion in 1995. Thus, the bank derivatives dealers demonstrated dominance in total derivatives volume of trading transactions. The GAO Report on Financial Derivatives, for example, noted that during the two-year period, 1994 and 1995, there were numerous entities that experienced substantial losses connected with the volume of trading of over-the-counter (OTC) derivatives and related financial products. In this respect, the losses were of a magnitude so severe that entities filed for bankruptcy, collapsed, or failed. such as Orange County, California, which was responsible for resources that amounted to $7.5 billion that covered 187 separate governmental agencies and, in December 1994, filed for bankruptcy after experiencing catastrophic losses in the amount of $1.7 billion based on the county treasurer's large and highly leveraged investments.[5]

Attention was placed on the large losses when, in February 1995, Baring Brothers & Co., Ltd, which was a British investment bank, collapsed due to more than $1 billion in futures and options trading losses on exchanges in Singapore and Japan, confirmed by the British Board of Banking Supervision inquiry.[6]

In both of its reports, which were published in 1994 and 1996 respectively, the GAO recommendations expressed uniformity in its focus to improve corporate governance and internal controls for major derivatives dealers and end users, close regulatory gaps, and establish comprehensive and consistent accounting standards internationally, which have yet, in 1996, to be fully implemented.[7]

There were special types of derivatives losses that gained significant recognition by 1994, and they were specific types of mortgage-backed securities known more appropriately as collateralized mortgage obligations (CMOs); this is a term that is interchangeable with real estate mortgage investment conduits.

The mortgage-backed security is defined as financial products whose source of payments generate from, or are driven by, a selected pool of home mortgages. The producers of multiclass mortgage-backed securities, by the rearrangement as well as the restructuring of the classification of the mortgage payments, have the capability to originate securities that emphasize yield, risk, and maturities.

Therefore, the payments to investors by issuers of multiclass mortgage-backed securities might be divided into the classification of securities by principal-only and interest-only securities that permit both to be sold separately. The worth of these two latter securities relate directly to mortgage prepayment rates that are sensitive to changing interest rates under the provision of two government-sponsored enterprises (GSEs), i.e., (a) the Federal National Mortgage Association, or Fannie Mae, and (b) the Federal Home Loan Mortgage Corporation, or Freddie Mac, that in fact have dominance over the issuance of multiclass mortgage-backed securities.[8]

An example of losses incurred by CMOs is the case of Capital Corporate Federal Credit Union, which was better known as Cap Corp, formerly one of the largest U.S. corporate credit unions. Cap Corp's losses totaled $61 million in February 1995. The National Credit Union Administration placed Cap Corp into conservatorship in January 1995 because it could not meet its financial obligations to its members.[9]

THE LONG-TERM CAPITAL MANAGEMENT (LTCM) NEAR CRASH

In the "Report of the President's Working Group on Financial Markets on Hedge Funds, Leverage, and the Lessons of Long-Term Capital Management (LTCM)," which was transmitted to Congress on April 28, 1999, the treatment of the term "hedge fund" is the embodiment of the descriptions of various differentiated forms of investment instruments that share common ties even though the term is not defined in the statutes. Hedge fund is a difficult and complex financial instrument.

The report of the President's Working Group on Financial Markets has defined the term "hedge fund" to mean the inclusion of any pooled investment vehicle that is privately organized, administered by professional investment managers, and not widely available to the public.[10] The President's Working Group report continued and noted that the primary investors in hedge funds are wealthy individuals and institutional investors. Moreover, hedge fund managers usually invest in the funds that they manage. The entities which are classified in categories that performed as hedge funds, in fact, are frequently formed and organized as limited partnerships or limited liability companies and, in several cases, have been domiciled outside the United States.[11]

The growth, expansion, and near crash of Long-Term Capital Management is an epic of enormous proportions, similar to a "poison pawn" strategy move, between January and September 1998 when LTCM lost almost 90 percent of its capital. Long-Term Capital Management, L.P. originated in 1994 as a Delaware limited partnership with its main offices located in Connecticut, and the fund that it operated, known as Long-Term Capital Portfolio, L.P. (the LTCM Fund or the Fund), is a Cayman Islands partnership that was the investment vehicle for a number of feeder funds, which were structured to meet the tax, regulatory, or accounting concerns of different classes of investors from different countries.[12]

Long-Term Capital Management is the representation of a network of limited partnerships and limited liability companies. LTCM possesses an operational long-term capital portfolio or master fund/feeder fund organizational structure formed to invest its assets resulting thereby in assets of the feeder funds being invested by the master fund, which is Long-Term Capital Portfolio.[13]

A distinguishing characteristic and dimension of a structural determinant of market conduct and performance that involve companies such as Long-Term Capital Management and its financial rivals encompassed the high degree of seller concentration and product differentiation as well as the condition of entry among the established sellers in the banking and financial industry.

From the perspective of the number and size distribution and the magnitude of 1998 Notional Value of Total OTC Derivatives contracts presentation

Table 4.4. Comparison of Total Notional Value of Derivatives Contracts[14] (Dollars in Billions)

Entity	*Total Notional Value*
LTCM	$1,400
The Goldman Sachs Group, L.P.	$3,410
Lehman Brothers Holdings, Inc.	$2,398
Merrill Lynch & Co., Inc.	$3,470
Morgan Stanley Dean Witter & Co.	$2,860

in Table 4.4: Comparison of Total Notional Value of Derivatives Contracts, there are four major securities and futures firms larger than LTCM. The dominant firm in the OTC derivatives market is not LTCM. More specifically, and around June 1998, there existed an estimated 2,500 to 3,500 hedge funds that engaged in the management of around $200 billion to $300 billion in capital and $800 billion to $1 trillion in total assets. From the standpoint of size, the hedge funds are essentially small when comparison is made with commercial banks which had nearly four times as many assets while mutual funds had five times as many assets in excess of LTCM.[15]

The Bank for International Settlements (BIS) Total Notional Contract amounts of derivatives contracts outstanding, and detailed in Table Vector 02: Global Over-the-Counter (OTC) Derivatives Accumulation Position Market, were estimated to be $55.7 trillion at the end of March 1995; $80.3 trillion at the end of December 1998: and $88.2 trillion at the end of December 1999, thereby providing the representation of the volume of growth and expansion of global OTC derivatives on an international scale, which is a pattern that is in contrast to the near crash of LTCM.

In the deepening of the contrast to LTCM's performance, there were fifteen major U.S. over-the-counter (OTC) derivatives dealers, and they accounted for the volumes of derivatives that have increased each year from 1990 to 1995.[16]

An illustration of the formation of the spread of institutional and financial catastrophic consequences has origins in the following Table 4.5: Notional/ Contract Amounts of Derivatives Reported by the 15 Major U.S. OTC Derivatives Dealers Year-End 1990 Through Year-End 1995.

The list of the identified fifteen major U.S. OTC derivatives dealers that are included in Table 4.5 consists of The Chase Manhattan Corporation; Citicorp; J. P. Morgan & Co., Inc.; Bankers Trust New York Corporation; BankAmerica Corporation; NationsBank Corporation; First Chicago Corporation; The Goldman Sachs Group, L.P.; Salomon, Inc.; American International

Table 4.5. Notional/Contract Amounts of derivatives Reported by the 15 Major U. S. OTC Derivatives Dealers Year-End 1990 Through Year-End 1995[17] Dollars in Billions

Dealers	*1990*	*1991*	*1992*	*1993*	*1994*	*1995*
Banks (7)	$5,350	$5,811	$7,574	$10,353	$13,724	$15,809
Securities Firms (5)	$1,730	$2,188	$2,967	$4,474	$5,880	$6,966
Insurance Companies (3)	$193	$257	$403	$634	$798	$985
Total (15)	$7,273	$8,256	$10,944	$15,461	$20,402	$23,760

Group, Inc.; The Prudential Insurance Company of America; and General Reinsurance Corporation.

Within the fifteen major U.S. OTC derivatives dealers that are identified in Table Epsilon 05, the bank dealers provide dominance in total derivatives volume with an estimated 69 percent yearly from 1990; second, the securities companies share is represented with an estimated 27 percent followed by, third, the insurance companies that accounted for 4 percent.[18] The fifteen major U.S. OTC derivatives dealers did not experience the LTCM near-crash consequences with its 90 percent loss of capital.

Why did LTCM, acknowledged as an enormous hedge fund and pooled investment entity that is engaged in active trading of various types of securities and commodities and is privately organized, administered, and managed by professional investment managers, but not widely available to the public and not statutorily defined, in effect, experience near-crash performance results, near collapse, or near failure? Why did the competitive firms or rivals of LTCM,such as derivatives dealers, commercial banks, insurance companies, and investment banks, in fact continue their successful operational performance and not experience LTCM's near-crash results, near collapse, or near failure?

There are several answers and responses that have been offered and made available, and some are part of the presentation below. The author's assessment and direct response to the preceding two questions relies heavily on Long-Term Capital Management's near-crash performance and the domination and totality of LTCM's commitment to the origination and complete devotion to inefficient, inadequate, deceptive, and synthetic financial products underlying variables that estimate speculative over-the-counter (OTC) derivatives contracts, changes in pricing, and include the following structural components:

1. The derived value at risk (VAR) theoretical model with probabilities of ranking potential commodity value in the determination of changes in prices;
2. Standard & Poor's Index;
3. Reference rates and underlying estimates of changes in pricing;
4. Consumer Price Index;
5. Nobel Prize in Economics Sciences winners Professor Robert Merton and Professor Myron Scholes's origination and utilization of an insufficient pure competition market formula for pricing commodity, event, options on stock, credit default swaps, and securities, and articles published in the *Journal of Political Economy* in 1973[19] that targeted estimates of changes in pricing the underlying variables that comprise OTC derivatives;

6. Interest rates fluctuations;
7. Foreign currency exchange rates variations;
8. Index reflects the estimated collective value of various financial products, or assets, i.e. stocks, bonds, and commodities;
9. Over-the-counter (OTC) derivatives huge volume of trading derived from estimated financial variables whose value is determined from and driven by flawed pure competition economic methods and insufficient underlying sources and unreal components, computer pricing models, indexes events, weather reports, interest rates, and foreign currency exchange rates that are not inclusive in the U.S. Money Stock Measures M1 and M2; the impact force resulting thereby is the formation of the near crash of Long-Term Capital Management.

Over-the-counter (OTC) derivatives are not U.S. dollars. LTCM's major failure occurred in its fallacy in its accumulation and utilization of the huge volume of trading in OTC derivatives as a replacement and substitute for U.S. dollars assets, in effect, trading in global OTC derivatives accumulation without corresponding U.S. dollar trading.

The money stock measures of the United States that are authorized and published by the Board of Governors of the Federal Reserve System are included is the following definition from

H.6 (508) "Money Stock Measures, M1 and M2":

1. M1 consists of (1) currency outside the U.S. Treasury, Federal Reserve Banks, and the vaults of depository institutions; (2) traveler's checks of nonbank issuers; (3) demand deposits at commercial banks (excluding those amounts held by depository institutions, the U.S. government, and foreign banks and official institutions) less cash items in the process of collection and Federal Reserve float; and (4) other checkable deposits (OCDs), consisting of negotiable order of withdrawal (NOW) and automatic transfer service (ATS) accounts at depository institutions, credit unions share draft accounts, and demand deposits at thrift institutions. Seasonally adjusted M1 is constructed by summing currency, traveler's checks, demand deposits, and OCDs, each seasonally adjusted separately.
2. M2 consists of M1 plus (1) savings deposits (including money market deposit accounts); (2) small-denomination time deposits (time deposits in amounts of less than $100,000), less individual retirement account (IRA) and Keogh balances at depository institutions; and (3) balances in retail money market mutual funds, less IRA and Keogh balances at money market mutual funds. Seasonally adjusted M2 is constructed by

summing savings deposits, small-denomination time deposits, and retail money funds, each seasonally adjusted separately, and adding this result to seasonally adjusted M1.[20]

The near-crash crisis in LTCM's management is rooted in its inefficient managerial organization and executive decision making, and the vast disparity between the low level base of dollar balance on its balance sheet and the corresponding gigantic volume of trading in global OTC derivatives accumulation which, in effect, is the representation of the condition of de facto bankruptcy thereby triggering the near crash and an emergency demand for huge increases in U.S. dollars as required assets.

In September 1998 the Federal Reserve Bank of New York carried out an intervention and took decisive action through its orchestration of a $3.6 billion recapitalization of Long-Term Capital Management by fourteen major over-the-counter derivatives dealers in response to LTCM's amassed in excess of $1 trillion in notional amount of OTC derivatives along with $125 billion of securities on $4.8 billion of capital without the knowledge of its major derivatives counterparties or federal regulators.[21]

There are other explanations for the near collapse of LTCM. For example, the U.S. General Accounting Office (GAO) Report to Congressional Requesters, *Long-Term Capital Management: Regulators Need to Focus Greater Attention on Systemic Risk*, GAO/GGD-00-3, dated October 1999, in fact structured its own assessment of LTCM's near collapse, its recapitalization, and how its positions became large and leveraged enough to be deemed a potential systemic threat.

This GAO LTCM report consists of a detailed-knowledge account and chronology of events that recognize LTCM's successful achievement in the establishment of leveraged trading positions of a magnitude and level that raised the threat of systemic risk, due largely to the banks and securities and futures companies that were its creditors and counterparties that failed to enforce their own risk management standards.[22]

The GAO report announced its broadly defined definition of leverage to mean the ratio between some measure of risk and capital. The simple balance sheet leverage is defined as assets divided by capital. However, this measure of leverage does not take into account the risk from off-balance sheet activities, such as the utilization of derivatives.

The issue of systemic risk, as referenced above, is generally defined as the risk that a disruption (at a firm, in a market segment, to a settlement system, etc.) could cause widespread difficulties at other firms, in other market segments, or in the financial system as a whole.[23]

This near-collapse review by the GAO Report on Long-Term Capital Management, GAO/GGD-00-3, especially in its treatment of the recapitalization decision, demonstrated that on September 23, 1998, there were fourteen of the world's largest banks and securities companies that agreed to the recapitalization of Long-Term Capital Management to avoid its disorderly liquidation, and the companies entered an agreement with LTCM on September 28, 1998, just five days later.

In taking action to avoid LTCM's near collapse, the fourteen banks and securities companies made a contribution of $3.6 billion (which represented 90 percent of the fund's net asset value at the time) to the fund through a new investment vehicle and general partner called Oversight Partner 1. The formation of this organization and the consortium members' affiliations are bound to the following institutions:[24]

1. Barclays PLC
2. Bankers Trust Corporation
3. Chase Manhattan Corporation
4. Credit Suisse First Boston Corporation
5. Deutsche Bank AG
6. Goldman Sachs Group
7. Lehman Brothers Holdings, Inc.
8. Merrill Lynch & Co., Inc.
9. J. P. Morgan & Company, Inc.
10. Morgan Stanley Dean Witter & Co.
11. Paribus
12. Societe Generale
13. Travelers Group Inc.
14. UBS AG

Oversight Partner 1 was approved and granted the general authority for management decision making and operations of the fund. This consortium formed an oversight committee which consisted of six members, namely, UBS; J. P. Morgan; Morgan Stanley; Goldman Sachs; Salomon Smith Barney; and Merrill Lynch.

The near crash of LTCM has a different outcome from that of the author's assessment. LTCM's near collapse focused on investment opportunities that involved global fixed-income trading strategies; it purchased bonds that it considered undervalued and sold bonds that it considered overvalued—fixed-income arbitrage funds. The failed investment strategy of LTCM involved taking long and offsetting short positions in securities, for example, buying an "old" two-year U.S. Treasury and selling (shorting) a corresponding "new" two-year U.S. Treasury.

The near-collapse investment strategy conducted by LTCM involved long and short positions in securities that would be "convergence trading" if specific models were valid. For example, conditional on the pricing model being correct, the positions in the securities would satisfy the conditions of arbitrage—for example, writing call options on a stock by a dynamic trading strategy in the underlying stock that should exactly replicate the payoff on the calls if the option-pricing model is correct. Another example would be dynamic trading in U.S. Treasury–based futures and securities that should produce arbitrage profits if the underlying model of interest rate dynamics and the term structure are valid.

LTCM's near-collapse investments were largely made up of bond arbitrage, for example, primarily arbitraging the difference between U.S. Treasuries with the same maturity but different issuance dates. The investment objective of LTCM's near collapse was the maximization of the expected total return on its portfolios on a risk-adjusted basis by applying pricing and analytical models and undertaking proprietary trading on a leveraged basis.

According to the GAO Long-Term Capital Management Report, GAO/GGD-00-3, the prevailing market conditions became the catalyst for LTCM's near collapse. This catalyst was initiated by the announcement issued from the Russian government on rescheduling payments on some of its debt obligations and imposing a moratorium on payments by Russian banks on certain obligations and sent global markets into a tailspin. The result of the Russian default was a dramatic increase in credit spreads and decrease in liquidity. Investors responded with a "flight to equality" and liquidity. The overall effect on Long-Term Capital Management's near collapse resulted in LTCM experiencing losses on positions in numerous markets. Then, on Sunday, September 20, 1998, a team that included officials from the Federal Reserve Bank of New York and the U.S. Treasury met with near-collapse LTCM's partners and learned the broad outlines of LTCM's major positions in credit and equity markets, and how the positions were deteriorating. The team realized the impact LTCM's position was having on markets around the world.

The next day, September 21, 1998, the Federal Reserve Bank of New York contacted three of LTCM's largest creditors; these calls led to the creation of the consortium of fourteen commercial banks and securities firms that recapitalized the fund. On September 23, 1998, the fourteen banks and securities firms agreed to recapitalize LTCM to avoid its disorderly liquidation. On September 28, 1998, they entered into an agreement with LTCM. The fourteen banks and securities firms made a contribution of about $3.6 billion to the fund through a new investment vehicle,generally called Oversight Partner 1.

NOTES

1. TABLE 4.1a-b: Global OTC Drivatives Accumulation Positions Market is compiled from the Bank for International Settlements (BS) numerous published sources among which are included BIS Press Releases, "Global OYC Derivatives Market atend-December 1999 . . . 2008" (http://w.w.w.bis.org/press/p990602.htl) BIS Quarterly Review, June 2012, Table 19' 2009, 2010 and 2011; U. S. General Accounting Office (GAO) Report to Congressional Committees, FINANCIAL DERIVATIVES Actions Taken or proposed Since May 1994, (GAO/GGD/AIMD-97-8), November 1996, p. 3 (for year 1995); Bank for International Settlements Moneteary and Economic Department, Triennial and Semiannual Surveys Positions in Global Over-the-Counter (OTC) Derivatives Markets at end-June 2010, Basil, Switzerlnd, November 20010; and Bank for International Settlments, Triennial Central Bank Survey, Foreign Exchange and Derivtives Market Activity in 2007, Basil, Switzerland, December 2007.
2. United Styates General Accounting Office Report to Congressional Committees, FINANCIAL DERIVATIVES Actions Taken or Proposed Since May 1994, GAO/GGD/AIMD-97-8
3. Ibid., p. 27
4. Ibid.
5. Ibid., Appendix-I, pp. 122–125
6. Ibid., Appendix-I, pp. 126–127
7. Ibid., p. 15
8. Ibid., pp. 24–26
9. Ibid., p. 125
10. Report of the President's Working Group on Financial Markets on Hedge Funds, Leverage, and the Lessons of Long-Term Capital Management, Department of the Treasury, Board of Governors of the Federal Reserve System, Securities and Exchange Commission, Commodity Futures Trading Commission, Washington, DC, April 1999, p. 1
11. Ibid.
12. Ibid., p. 10
13. U. S. General Accounting Office (GAO) Report to Congressional Requesters Long-Term Capital Management, Regulators Need to Focus Greater Attention on Systemic Risk, GAO/GGD-00-3, Washington, DC, October 1999, p. 38
14. Ibid., p. 25
15. Ibid., p. 38
16. Ibid., pp. 26–27
17. Ibid., pp. 27, 156–158
18. Ibid., p. 27

19. Nicholas Dunbar, The Devil's Derivatives, The Untold Story of the Slick Traders and Hapless Regulators Who Almost Blew Up Wall Street . . . and Are Ready to Do It Again, Boston, MA: Harvard Business Review Press, 2011, Chapter 2, pp. 31–39

20. Federal Reserve Statistical Release, H.6. Money Stock Measures, Release Date: August 23, 2012, H.6. (508) Table 1 Money STock Measures (www.federalreserve.gov/releases/h6/current/h6.htm)

21. The Financial Crisis Inquiry Report, Final Report of the National Commission on the Causes of the Financial and Economic Crisis in the United States, Official Government Edition, Submitted by the Financial Crisis Inquiry Commission Pursuant to Public Law 111–21, January 2011, Washington, DC: U. S. Government Printing Office, pp. 47–48

22. GAO Report to Congressional Requesters Long-Term Capital Management, Regulators Need to Focus Greater Attention on Systemic Risk,, GAO/GGD-00-3, Washington, DC , October 1999, p. 1

23. Ibid., pp. 1–2

24. Ibid., pp. 44–45 Staff Report Prepared by the Permanent Subcommittee on Investigations of the Committee on Homeland Security and Governmental Affairs United States Senate, THE ROLE OF MARKT SPECULATION IN RIISING OIL AND GAS PRICES: A Need to Put the COP Back On the Beat, 109th Cong., 2nd Sess, June 27, 2006, Washington, DC: Government Printing Office, 2006, pp. 1–7 and pp. 28–37

5

The Financial Crisis, Corporations, Concentration of Control and Market Power

The Sherman Act, 15 USC Section 1: "Every contract, combination in the form of trust or otherwise, or conspiracy, in restraint of trade or commerce among the several states, or with foreign nations, is declared to be illegal. Every person who shall make any contract or engage in any combination or conspiracy hereby declared to be illegal shall be deemed guilty of felony, and on conviction thereof, shall be punished by fine not exceeding $100,000,000 if a corporation, or, if any other person, $1,000,000, or by imprisonment not exceeding 10 years, or by both said punishments, in the discretion of the court."

The Sherman Act, 15 USC Section 2: "Every person who shall monopolize, or attempt to monopolize, or combine or conspire with any other person, to monopolize any part of the trade or commerce among the several states, or with foreign nations, shall be deemed guilty of a felony, and, on conviction thereof, shall be punished by fine not exceeding $100,000,000 if a corporation, or, if any other person, $1,000,000, or by imprisonment not exceeding 10 years, or by both said punishments, in the discretion of the court."

The financial crisis and complexities of 2007 and 2008 are rooted in the current and historical market structure dominance characteristics, performance, and the "money trust" under the direction of a few firms. At the center of the market structure dominance characteristics, performance, and the "money trust" is the constituted and unleashed domination, control, and consolidation of the processes of concentration that are accompanied by the increased concentration of control of money and credit.[1]

The definition of "money trust" is, as the Report of the Pujo Committee states, an established and well-defined identity and community of interest between a few leaders of finance that has been created and is held together through stock holdings, interlocking directorates, and other forms of domination over banks, trust companies, railroads, public service and industrial corporations and that has resulted in a vast and growing concentration of control of money and credit in the hands of a comparatively few men.

PROCESSES OF CONCENTRATION

Essentially, the linchpin of the processes of concentration and the increased concentration of control of money and credit, according to the *Report of the Committee Appointed Pursuant to House Resolutions 429 and 504 to Investigate the Concentration of Control of Money and Credit*, February 28, 1913, (also known as the Report of the Pujo Committee) takes the form of and is driven by, first, the completion of mergers, acquisitions, and consolidations of competitive banks and trust companies and, at the same time, strengthening of the centralization of management.[2]

Second, having the objective of acquiring control and domination, the same completion of mergers, acquisitions, and consolidations of competing banks and trust companies and strengthening of the concentration of management interests, in fact, directly become major stockholders in potentially competitive banks and trust companies.Third, the concentration of control of money and credit has been developed through the integration of potentially competitive banks and trust companies as a result of interlocking directorates.

Fourth, through the same domination structure, the processes of concentration have been realized through the influence that the more powerful banking houses, banks, and trust companies have secured in the management of insurance companies, railroads, producing and trading corporations, and public utility corporations by means of stock holdings, voting trusts, fiscal agency contracts, or representation upon their boards of directors, or through supplying the money requirements of railway, industrial, and public utilities corporations and thereby being enabled to participate in the determination of their financial and business policies.[3]

Fifth, the increased concentration of control of money and credit has advanced through partnerships or the joint account arrangements between a few of the leading banking houses, banks, and trust companies in the purchase of security issues of the great interstate corporations, accompanied by the understandings of recent growth and, thus, increasing unfavorable restraint of trade practices or reduction of competition between such banking houses, banks, and trust companies engaged in the competition for business or in the purchase and sale of large issues of such securities.[4]

The preceding processes of concentration and increased concentration of control of money and credit produced their own originators, institutions, and agents of power and control who are responsible for the formation, advancement, and management of this concentration of control of money and credit conducted through one or a combination of the processes of concentration presented above and included principally:

- J. P. Morgan & Co.
- First National Bank of New York
- National City Bank of New York
- Lee, Higginson & Co. of Boston and New York
- Kidder, Peabody & Co. of Boston and New York
- Kuhn, Loeb & Co.

The 2007 and 2008 financial crisis and its dominant mega market power structure and multinational enterprises behavior and performance are, in fact, a representation and extension of J. P. Morgan & Co. and Morgan & Co., a combination known as the "House of Morgan," which is the principal in the processes of concentration that are assessed and critically presented throughout the *Report of the Committee Appointed Pursuant to House Resolutions 429 and 504 to Investigate the Concentration of Control of Money and Credit* (also known as the Pujo Committee Report).

As the dominant firm in the processes of concentration of control of money and credit, J. P. Morgan & Co. of New York and Morgan & Co. (the House of Morgan) and Drexel & Co. of Philadelphia, Pennsylvania, are one and the same:[5]

J. P. Morgan & Co. is a partner in the London banking house of J. S. Morgan & Co. and Harjes & Co. (the Paris House of Morgan); Morgan & Co. selected the entire board of directors of the Bankers Trust Co.; in affiliation with the Guaranty Trust Co., Morgan & Co. controlled the Guaranty Trust and Bankers Trust companies through voting trusts; in affiliation with National Bank of Commerce, J. P. Morgan Jr. as well as twelve directors of the Guaranty Trust Co. that is controlled by Morgan & Co. are directors of this bank. In the summary of affiliations with financial corporations, Morgan & Co. and their nominees, therefore, control or have a powerful influence in banks and trust companies in the city of New York.

Morgan & Co. was one of the organizers of the General Electric Company and is a stockholder and a depositary of the company and markets its security issues; Morgan & Co. organized International Harvester Co. and is a stockholder that marketed for the company; Morgan & Co.'s affiliation with the American Telephone & Telegraph Co. includes directors and being a stockholder that has, since 1906, procured financing in joint account with affiliated banks and bankers and marketed issues of securities for the New York Telephone Co. subsidiary, the Chicago Telephone Co. subsidiary, and the Pacific Telephone & Telegraph Co. subsidiary.

In affiliation with Interborough Rapid Transit Co., Morgan & Co. were its bankers and the head of a syndicate to purchase an impending bond issue; the Hudson & Manhattan Co. that owned and operated the tunnels under the Hudson River that connected New York and New Jersey was affiliated with Morgan & Co., which procured the marketing of its security issues and was itself a large holder thereof.

The 1913 Report of the Pujo Committee found that the First National Bank of New York was organized in 1863 by Morgan & Co. principals and partners. The directors were Morgan & Co. partners and J. P. Morgan that formed the majority of the executive committee; Morgan & Co. Partners controlled and dominated the management. The First National Bank controlled Chase National Bank. Two of the partners who also were two of the most active members of Morgan & Co. were vice presidents of the First National Bank while remaining directors. After the head of the First National Bank, Morgan & Co. is the largest stockholder in First National Bank, according to the Pujo Committee Report of 1913.

Furthermore, the First Security Co. was organized in February 1908 under the laws of the state of New York with a capital stock of $10 million by the stockholders of the First National Bank, the Report of the Pujo Committee continued, who authorized the officers of the bank to subscribe for and hold the stock of the new company as trustees under an agreement requiring that the stock of the security company shall always be owned by the same persons who own the stock of the bank and in the same proportions, and that no person not a director of the bank shall be a director of the security company. The trustees shall always be the president, vice presidents, and cashier of the bank.[6]

The Morgan & Co. principals and partners, in affiliation with the Chase National Bank, represented five of the nine directors of the Chase National Bank.

A Morgan & Co. principal, who took action prior to 1908, acquired in the interest of First National Bank a majority of the stock of the Chase Bank, transferring it in that year to the First Security Co., an arm of the First National Bank.

The bold, consolidated, and close cooperation that existed between Morgan & Co. and First National Bank was dominated by their joint purchases and underwriting of corporate securities. Morgan & Co. and First National Bank, through joint agreements, joint stock holdings, interlocking directorates, partnership transitions, and other relations were locked together, the 1913 Report of the Pujo Committee affirmed, in a complete and enduring community of interest.

The acquisition by Morgan & Co. of a large block of stock of National City Bank, with representation on its board of directors in which these two institutions along with First National Bank were joined, demonstrated a unison of interest and a continuity of cooperation between Morgan & Co., National City Bank, and First National Bank.

With specific reference to banking resources, the combined power of Morgan & Co., First National Bank, National City Bank, and First Security Co., through stockholdings, voting trusts, interlocking directorates, and other relations, has become the absolutely dominant factor in some cases and, at the same time, the most important single factor in the control of the following banks and trust companies in the city of New York:[7]

1. Bankers Trust Co.
2. Guaranty Trust Co.
3. Astor Trust Co.
4. National Bank of Commerce
5. Liberty National Bank
6. Chase National Bank
7. Farmers Loan & Trust Co.

The significance and magnitude of interlocking directorates gained enormous stature as one dominant market structure component of the processes of concentration and its documentation by the report, *Money Trust Investigation: Investigation of Financial and Monetary Conditions in the United States Under House Resolutions Nos. 129 and 504 Before a Subcommittee of the Committee on Banking and Currency: Interlocking Directorates*, Washington, DC: U.S. Government Printing Office, 1918.[8]

The conduct and behavior framework driven by the financial and economic crisis and collapse of 2007 and 2008 bears close resemblance to the past and core community of interest practices and the following frame-of-reference affiliations. First, for example J. P. Morgan & Co., First National Bank, National City Bank, the Guaranty Trust Co., and the Bankers Trust Co. together have 118 directorships in 34 bank and trust companies; 30 directorships in 10 insurance companies; 105 directorships in 32 transportation systems; 63 directorships in 24 producing and trading corporations; 25 directorships in 12 public utility corporations; in all, 341 directorships in 112 corporations.

Second, J. P. Morgan & Co., the Guaranty Trust Co., the Bankers Trust Co., and First National Bank together have 89 directorships in such banks and trust companies; 29 directorships in such insurance companies; 78 directorships in such transportation companies; 78 directorships in such transportation systems; 49 directorships in such producing and trading corporations; and 16 directorships in such public utility corporations; in all, 201 directorships.

Third, J. P. Morgan & Co., the Guaranty Trust Co., and the Bankers Trust Co. together have 78 directorships in such banks and trust companies; 29 directorships in such insurance companies; 64 directorships in such transportation systems; 44 directorships in such producing and trading corporations, and 14 directorships in such public utility corporations; in all, 229 directorships.

THE FINANCIAL AND ECONOMIC CRISIS AND COLLAPSE, 2007 AND 2008, GLOBAL OTC DERIVATIVES GIGANTIC ACCUMULATION AND U.S. MONEY STOCK DEMAND

What was the central reason for the financial and economic crisis and collapse of 2007 and 2008? It is the totality of the dominant concentration of control of economic power mobilization through the following: the voting rights in major corporations, the structure of corporate concentration, the interlocking directorates among the major corporations, and the processes of concentration of control—what the Report of the Pujo Committee of 1913 called the "Money Trust" that, in effect, established the global over-the-counter (OTC) derivatives' huge accumulation as the highest preference and priority, which resulted in a profound substitution, replacement, or declining utilization of demand for volumes of transactions and the holding of assets balances in U.S. Money Stock Measures, M1 and M2, the definition of which include the following (Federal Reserve Statistical Release, H.6. "Money Stock Measure," August 23, 2012):

1. M1 consists of (1) currency outside the U.S. Treasury, Federal Reserve Banks, and the vaults of depository institutions; (2) traveler's checks of nonbank issuers; (3) demand deposits of commercial banks (excluding those amounts held by depository institutions, the U.S. government, and foreign banks and official institutions) less cash items in the process of collection and Federal Reserve float; and (4) other checkable deposits (OCDs), consisting of negotiable order of withdrawal (NOW) and automatic transfer service (ATS) accounts at depository institutions, credit union share draft accounts, and demand deposits at thrift institutions. The seasonally adjusted M1 is constructed by summing currency, trav-

eler's checks, demand deposits, and OCDs, each seasonally adjusted separately.

2. M2 consists of M1 plus (1) savings deposits (including money market deposit accounts); (2) small-denomination time deposits (time deposits in amounts of less than $100,000), less individual retirement account (IRA) and Keogh balances at depository institutions; and (3) balances in retrial money market mutual funds, less IRA and Keogh balances at money market mutual funds. Seasonally adjusted M2 is constructed by summing savings deposits, small-denomination time deposits, and retail money funds, each seasonally adjusted separately, and adding this result to seasonally adjusted M1 (www.federalreserve,go/release/H6/current/H6.htm). This continued lodestone assessment by the author adds emphasis and places significance on this trade-off that encompasses the sacrifice or giving up of the greatest agglomeration and aggrandizement of the concentration of control of economic power and corporation ownership and control in exchange for the accumulation of global over-the-counter derivatives. This pattern of processes of the concentration, corporation ownership and partnerships, trade, commerce, and conduct, in fact, soared to the leadership framework trade-off and the central reason as the answer to the question presented—what was the central reason for the financial and economic crisis and collapse of 2007 and 2008?—this is the author's assessment.

The global over-the-counter derivatives position market, according to the Bank for International Settlement (BIS) has ranged from $270 trillion in 2005; $370 trillion in 2006; $507.9 trillion in 2007; $683.7 trillion in 2008; $605.0 trillion in 2009; $582.7 trillion in 2010; and $706.8 trillion in 2011. This accumulation of over-the-counter derivatives driven by the U.S. banking and financial system necessitated the action taken; the amount of money, the total guarantees, and lending limits that the Federal Reserve parceled out and committed has amounted to $7.7 trillion as of March 2009 for the purpose of rescuing the U.S. financial system. *Bloomberg Markets* magazine reported that the preceding $7.7 trillion as of March 2009 was more than half the value of everything produced in the U.S. that year.[9]

THE FINANCIAL CRISIS INQUIRY REPORT

The *Final Report of the National Commission on the Causes of the Financial and Economic Crisis in the United States*, a document submitted by the Financial Crisis Inquiry Commission (FCIC), delivered its cornerstone assessment in both "Crisis and Panic" and "The Growth of Derivatives," a most significant event in finance and important response to the question presented above—what is the cause of the financial and economic crisis and collapse of 2007 and 2008?

First, the Report of the Financial Crisis Inquiry Commission (FCIC), through its assessment of crisis and panic findings, places emphasis on the interconnection and elevation of three important events—September 15, 2008, the date of the bankruptcy of Lehman Brothers; the takeover of Merrill Lynch; followed within twenty-four hours by the rescue of AIG—marked the beginning of the worst market disruption in postwar American history and, simultaneously, an extraordinary rush to the safest possible investments.[10]

The FCIC Report found that, in the middle of the failure of Lehman Brothers, September 15, 2007, Morgan Stanley CEO John Mack told the FCIC that similar institutions have experienced a classic run on the bank while, at the same time, investors lost confidence in financial institutions, and the entire investment banking business model was under siege.[11]

CEO Jamie Diamond of J. P. Morgan sealed the position advocated by Morgan Stanley when his remarks made to the FCIC specified that the markets were very bad—the volatility, the illiquidity—some things could not trade at all. The U.S. unemployment level could reach an estimated 20 percent.[12]

The Report of the FCIC, through the independent and separate results of the study findings, has produced an additional response to the question, what is the cause of the financial and economic crisis and collapse of 2007 and 2008 in the United States; this added FCIC Report response to the question is found in the FCIC's "Growth of Derivatives: By Far the Most Significant Event in Finance During the Past Decade."

The FCIC Report's definition of derivatives is that derivatives are financial contracts whose prices are determined by or derived from the value of some underlying asset, rate, index, or event; they are not used for capital formation or investment, as are securities; they are instruments for hedging business risks or for speculating on changes in prices, interest rates, and the like.

Derivatives come in many forms. The most common form of derivatives is the over-the-counter (OTC) swaps and exchange-traded futures and options. Derivatives may be based on commodities (including agricultural products, metals, and energy products), interest rates, currency rate, stocks and indexes, and credit risk; they could even be tied to events such as hurricanes or announcements by government figures.[13]

For instance, OTC derivatives are traded by the large financial institutions such as bank holding companies and investment banks that act as derivatives dealers who are buying and selling contracts with customers; the OTC derivatives market is neither centralized nor regulated.

Most importantly, the Report of the FCIC emphasized the position advocated by U.S. Treasury Secretary Robert Rubin, Federal Reserve Board Chairman Alan Greenspan, Securities and Exchange Commission Chairman Arthur Levitt, and Deputy U.S. Treasury Secretary Lawrence Summers, who viewed with favor and certainty that regulation of derivatives contracts and transactions that are privately negotiated by professionals is unnecessary. A moratorium was proposed on the Commodity and Futures Trading Commission's (CFTC) ability to regulate over-the-counter derivatives.

In addition, the Commodity Futures Modernization Act of 2000 (CFMA) was passed by Congress and signed by President Bill Clinton in December 2000; the CFMA, in effect, deregulated the OTC derivatives market and eliminated oversight by both the CFTC and Securities and Exchange Commission (SEC). The CFMA provided protection of over-the-counter derivatives from nearly all regulation or oversight.

A significant volume of OTC derivatives during the financial crisis of 2007 and 2008 was the credit default swap (CDS), which offered the seller a little potential upside at the relatively small risk of a potentially large downside. For example, the FCIC Report noted that the buyer of a credit default swap transferred to the seller the default risk of an underlying debt (but the seller does not own the CDS; the debt security could be any bond or loan obligation). The author's assessment is that the CDS contract volume of transactions constitutes an unreasonable restraint of trade and is in violation of the Sherman Act, Section 1 because the seller acts in a manner of offering insurance against a default in the underlying variable that is the base for the determination of the price of the OTC derivatives while the buyer owns the commodity, not the seller, and the result is that the buyer is liable for the default damage and, thereby, the subsequent elimination of the buyer and reduction in competitive market conditions. The seller can insure only what is owned by the seller and not what is owned or purchased by the buyer. Seller's interest is required for seller's insurance.

The exclusive dealing, the high degree of concentration, and market dominance along with the substantial risk of credit default swaps and other OTC derivatives, in fact, were dominated by a few of the very largest banks, investment banks, and insurance companies that controlled the dealings in OTC derivatives.

In the Report of the FCIC the concentration of control of the very largest banks and investment banks in the domination of derivatives in the past decade include, among U.S. bank holding companies, the 97 percent of the notional amount of OTC derivatives; millions of contracts were traded by just five large institutions in 2008 during the financial crisis:[14]

1. JPMorgan Chase
2. Citigroup
3. Bank of America
4. Wachovia
5. HSBC

In addition, the five largest investment banks are included among the world's largest OTC derivatives dealers.[15]

The question, what was the cause of the financial and economic crisis and collapse during 2007 and 2008?, was addressed by Chairman of the Board of Governors of the Federal Reserve System Ben S. Bernanke in the delivery of his testimony and statement, "Causes of the Recent Financial and Economic Crisis," before the FCIC on September 2, 2010. Chairman Bernanke announced that it is essential to distinguish between (1) triggers that he defined as the particular events or factors that touched off the crisis, and (2) vulnerabilities that he defined as the structural weaknesses in the financial system, as well as in the regulation and supervision that propagated and amplified the initial shocks.[16]

In the delivery of his testimony, Chairman Bernanke emphasized that the most prominent development that triggered the crisis was the prospect of significant loses on residential mortgage loans to subprime borrowers that became apparent shortly after home prices began to decline. In contrast to this direct testimony, Chairman Bernanke delivered his own correction when he testified that with more than $1 trillion in subprime mortgages outstanding, the potential for losses on these loans was large, however, the prospective subprime losses were clearly not large enough on their own to account for the magnitude of the financial crisis.[17]

Chairman Ben Bernanke, in the continuation of his direct testimony before the FCIC, stated that the daily movements in global equity markets, in fact, regularly impose aggregate gains or losses that are equal to or greater than all the subprime mortgage losses incurred at that time. The principal explanations of why the crisis was so severe and had such devastating effects on the broader economy, Chairman Bernanke continued, were the system's vulnerabilities in combination with the gaps in the government's crisis-response toolkit.

A trigger of the financial crisis was an event described as a "sudden stop" in June 2007 that involved syndicated lending to large and risky corporate borrowers. Once the province of banks, the funding for these leveraged loans had migrated recently from banks to special-purpose vehicles that funded themselves by issuing collateralized loan obligations (CLO), which is a type of asset-backed security. The CLOs were bought by a variety of investors that included asset-backed commercial paper (ABCP) vehicles.

Chairman Bernanke declared that there were other triggers that helped start the cascade of events that became the financial crisis. His assertion that when the financial system is sufficiently vulnerable the result is that some of those vulnerabilities occur in the private sector; many key vulnerabilities related directly to private-sector practices.

There was one vulnerability known as the overreliance on unstable short-term funding. Chairman Bernanke singled out the shadow banks which he defined as financial entities that serve as intermediaries to channel savings into investments that include securitization vehicles, asset-backed commercial paper (ABCP) vehicles, money market funds, investment banks, mortgage companies, and a variety of other entities, and in effect, are part of the shadow banking system. The shadow banks do not include regulated depository institutions such as commercial banks, thrifts, and credit unions.

In the pre–financial crisis period, the shadow banking system had come to play a major role in global finance; therefore, they are a source of some important vulnerabilities. The dependence of shadow banks on short-term uninsured funds made them subject to runs. The fears of a run, Chairman Bernanke continued, have the potential to become at least partially self-fulfilling. An increase in the risk of a run induces financial firms to hoard liquidity.

Money market mutual funds proved particularly vulnerable to liquidity pressures while, at the same time, the currency mismatches also contributed to the disruption of wholesale funding patterns during the crisis. In his delivery of remarks before the Financial Crisis Inquiry Commission (FCIC), Chairman Bernanke stated that over-the-counter (OTC) derivatives had a mixed record in the crisis; he testified that throughout the period virtually all derivatives contracts settled according to their terms. At times the complexity and diversity of derivatives instruments also posed problems. The clearing and settlement infrastructure was seriously inadequate.

This final report of the FCIC differs substantially from the direct testimony delivered by Chairman Ben Bernanke, specifically with reference to the assessment of over-the-counter derivatives and the role of the Federal Reserve System during the financial crisis.

In its conclusion the report of the Financial Crisis Inquiry Commission found that the financial crisis was avoidable. The leaders of finance and industry and the public stewards of the U.S. financial system, in effect, ignored warnings and failed to question, understand, and manage evolving risks.

For example, there was (a) an explosion in risky subprime lending and securitization; (b) an unsustainable rise in housing prices; (c) widespread reports of egregious and predatory lending practices; (d) dramatic increases in household mortgage debt and exponential growth in financial firms' trading activities; (e) unregulated derivatives and short-term "Repo" lending markets; and (f) pervasive permissiveness.

The prime example, the report of the FCIC concluded, is the Federal Reserve's pivotal failure to stem the flow of toxic mortgages, which it could have done by setting prudent mortgage-lending standards. The Federal Reserve was the one entity empowered to do so and the Federal Reserve did not, the report of the FCIC found.

There is another important source that has provided a response to the question, what is the cause of the financial and economic crisis and collapse of 2007 and 2008? This significant source is *Wall Street and the Financial Crisis: Anatomy of a Financial Collapse, Report and Appendix before the Permanent Subcommittee on Investigations of the Committee on Governmental Affairs*, U.S. Senate, 112th Congress, 1st Session, Volume 5, Part I, April 13, 2011. It was in November 2008 that the Permanent Subcommittee on Investigations of the U.S. Senate initiated its investigation into some of the key causes of the financial crisis. This report on Wall Street and the financial crisis makes an assessment on how four causative factors, in effect, fueled the 2007–2008 financial crisis, namely, (1) High Risk Lending: Case Study Washington Mutual Bank; (2) Regulatory Failure: Case Study of the Office of Thrift Supervision; (3) Inflated Credit Ratings: Case Study of Moody's and Standard & Poor's; and (4) Investment Bank Abuses: Case Study of Goldman Sachs and Deutsche Bank.

The assessment that is presented in the high-risk lending case study of Washington Mutual Bank, which is identified as one of four causes of the financial crisis by this report, in fact, affirmed how the Washington Mutual Bank (known as WaMu) strategy for growth and profit contributed to the origination and securitization of hundreds of billions of dollars in poor- quality mortgages that undermined the U.S. financial system.[18]

Washington Mutual Bank was a wholly owned subsidiary of its parent holding company, which is Washington Mutual, Inc. Between 1996 and 2002, the report on Wall Street and the financial crisis announced among its findings that WaMu acquired over a dozen other financial institutions that included American Savings Bank, Great Western Bank, Fleet Mortgage Corporation, Dime Bancorp, PNC Mortgage, and Long Beach Corp., thereby expanding to become the nation's largest thrift and its sixth-largest bank. In addition, WaMu became one of the largest issuers of home loans in the U.S.

After placing in reserve some of those U.S. home loans for its own investment portfolio, WaMu then sold the remainder of the home loans to Wall Street investors and to Fannie Mae or Freddie Mac. As a result of its increasing expansion and securitization from 2000 to 2008, WaMu sold over $500 billion in loans to Fannie Mae and Freddie Mac, and that represented more than one-quarter of its loan production during those years.[19]

This report of Wall Street and the financial crisis noted that the formal initiation of WaMu's high risk lending strategy dated from January 2005 with the presentation submitted to the board of directors for approval. Moreover, WaMu developed an accumulation of high risk home loan origination to borrowers who had traditionally qualified for them in 2004. WaMu adopted a high-risk strategy to issue high-risk mortgages and then mitigate some of that risk by selling or securitizing many of the home loans.

By 2007 the problem with WaMu's high-risk lending strategy worsened as delinquencies increased, the securitization market dried up, and the bank was unable to find buyers for its high- risk loans or related securities. On September 18, 2008, the Office of Thrift Supervision (OTS) and the Federal Deposit Insurance Corporation (FDIC) lowered Washington Mutual's rating to "4," which indicated that bank failure was a possibility. An estimated $17 billion in deposits left Washington Mutual at about the same time. Because of its liquidity problems and poor-quality assets, both the OTS and FDIC decided to close WaMu.

On September 25, 2008, the OTS seized Washington Mutual Bank, placed it into receivership, and appointed the FDIC as the receiver, according to this report of Wall Street and the financial crisis. The FDIC facilitated its immediate sale to J. P. Morgan Chase for $1.9 billion. Soon after, Washington Mutual, Inc., WaMu's parent, declared bankruptcy.[20]

This report has ranked as the second of four causes of the financial crisis the "Regulatory Failure: Case Study of the Office of Thrift Supervision (OTS)"; its presentation is what it called a justifiable charge of irresponsibility against the OTS.

The largest thrift under the supervision of the OTS, by late 2008, was Washington Mutual Bank, which with more than $300 billion in assets, $188 billion in deposits, over 200 branches located in fifteen states, 43,000 employees, was among the eight largest financial institutions that were insured by the FDIC. WaMu's collapse in September 2008 marked one of the most spectacular failures of federal bank regulators in recent history.[21]

The scathing criticism of WaMu's operation and performance by the report described how in the years leading up to the financial crisis the Office of Thrift Supervision failed to prevent Washington Mutual's increasing use of high-risk lending practices and its origination and sale of tens of billions of dollars in poor-quality home loans, in spite of the fact that federal bank regulators are charged with ensuring that U.S. financial institutions operate in a safe and sound manner.

The OTS examiners uncovered more than five hundred serious deficiencies in WaMu lending risk management and operations during a five-year period from 2004 to 2008, however, the OTS failed to take a public enforcement action against Washington Mutual.[22]

The U.S. Senate Permanent Subcommittee on Investigations held a hearing on April 16, 2010, at which the OTS, the FDIC, and the Inspector General (IG) officials provided testimony and some ninety-two exhibits were released; also released was the FDIC's and the Treasury IG's Joint Report on Washington Mutual. In connection with the hearing, the report *Wall Street and the Financial Crisis: Anatomy of a Financial Collapse*, in fact, included the U.S. Senate Permanent Subcommittee on Investigations Joint Memorandum from Chairman Carl Levin and Ranking Member Tom Coburn summarizing the investigation into the role of the regulators overseeing Washington Mutual.

One conclusion taken from the Levin-Coburn memorandum endorsed their view that Washington Mutual was just one of many financial institutions that federal banking regulators allowed to engage in such high-risk home loan lending practices that they resulted in bank failure and damage to financial markets. The ineffective role of bank regulators was a major contributor to the 2008 financial crisis that continued to afflict the U.S. and world economy today.[23]

The Levin-Coburn memorandum, which included its joint findings of fact referring specifically to the federal regulators' role in the Washington Mutual case, reaffirmed the following: (a) from 2003 to 2008 the Office of Thrift Supervision repeatedly identified important problems with Washington Mutual's lending practices, risk management, and asset quality, but failed to force adequate corrective action and resulted in the largest bank failure in U.S. history; (b) the OTS allowed Washington Mutual and its affiliate, the Long Beach Mortgage Company, to engage year after year in shoddy lending and securitization practices and failed to take enforcement action to stop its origination and sale of loans with fraudulent borrower information, appraisal problems, errors, and notoriously high rates of delinquency and loss; and (c) the OTS abdicated its responsibility to ensure the long-term safety and soundness of Washington Mutual by concluding that short-term profits obtained by the bank precluded enforcement action to stop the bank's use of shoddy lending and securitization practices and unsafe and unsound loans.[24]

The third of the four causes of the financial crisis is identified as "Inflated Credit Ratings: Case Study of Moody's and Standard & Poor's" and is included in the report on Wall Street and the financial crisis, which delivered an indictment of the following predatory practices and behavior. For example, Moody's Investors Service, Inc. (Moody's) and the Standard & Poor's Financial Services LLC (S&P), ranked as the two largest credit rating agencies (CRAs) in the United States, issued the AAA (triple-A) ratings that made residential mortgage-backed securities (RMBS) and collateralized debt obligations (CDOs) seem like safe investments, helped build an active market for these securities, and then, beginning in July 2007, downgraded the vast majority of those AAA ratings to junk status. The July 2007 mass downgrades sent the value of mortgage-related securities plummeting, precipitated the collapse of the RMBS and CDO secondary markets, and perhaps more than any other single event triggered the financial crisis.[25]

Credit ratings are supposed to provide independent assessments of the creditworthiness of specific financial instruments such as a corporate bond, mortgage-backed security, or CDO. The U.S. has three major credit rating agencies: Moody's, Standard & Poor's (S&P), and Fitch Ratings Ltd., each of which is a nationally recognized statistical rating organization (NRSRO).

In fact, these three credit rating agencies or firms issue an estimated 98 percent of total credit ratings and collect 90 percent of total credit rating revenue, thereby raising issues of unreasonable restraint of trade, monopolization, and exclusive dealing under the Sherman Act, Section 1 and Section 2. Under the Dodd-Frank Act the formation of the credit rating agency in the executive branch of the government, specifically concerning federal securities and instruments, U.S. Money Stock Measures M1 and M2 financing, and not the private-sector credit rating agencies involvement or inclusion, would avoid the pitfalls of the credit rating firms' role in the financial and economic crisis of 2007 and 2008—the author's government-sponsored enterprise (GSE) and multinational enterprise (MNE) strategy.

In 1970, the credit rating agencies changed to an "issuer-pays" model that means the party seeking to issue a financial instrument, such as a bond or security, pays the credit agency to analyze the credit risk and assign a credit rating to the financial instrument.

The report *Wall Street and the Financial Crisis: Anatomy of an Economic Collapse* findings exposed the mass credit rating downgrades predatory practices formed as Moody's and S&P together issued investment-grade ratings for tens of thousands of RMBS (or mortgage-backed security debt instrument secured by a pool of mortgages, whether residential or commercial) and CDO securities (collateralized debt obligation type of security often composed of the riskier portions of mortgage-backed securities), earning substantial sums for issuing these ratings.

In addition, in the mid-2007 period, both credit rating agencies, Moody's and S&P, suddenly reversed course and started downgrading hundreds, then thousands of RMBS and CDO ratings. These huge and mass downgrades shocked the financial markets, contributed to the collapse of the subprime RMBS and CDO secondary markets, triggered sales of assets that had lost investment grade status, and damaged holdings of financial firms worldwide. Perhaps more than any other single event, the report *Wall Street and the Financial Crisis: Anatomy of a Financial Collapse* makes the assertion that the sudden mass downgrades of RMBS and CDO ratings were the immediate trigger for the financial crisis.[26]

During the first half of 2007 the predatory practices conduct and behavior continued despite the news of failing subprime lenders and increasing subprime mortgage defaults. Moody's and S&P continued to issue AAA credit ratings for a large number of RMBS and CDO securities.

In the first week of July 2007 alone, S&P issued over 1,500 new RMBS ratings, which represents a number that almost equaled the average number of RMBS ratings that it issued in each of the preceding three months.[27]

The predatory pattern expanded from July 5 to July 11, 2007, when Moody's issued approximately 675 new RMBS ratings that nearly doubled its weekly average in the prior month. The timing of this surge of new ratings on the eve of the mass downgrades is troubling, and raises serious questions about whether S&P and Moody's quickly pushed the ratings through in order to avoid losing revenues before the mass downgrades began.[28]

The predatory practices and conduct of Moody's and S&P deepened in the second week of July 2007 as S&P and Moody's initiated the first of several mass ratings downgrades that shocked the financial markets. Moreover, S&P placed on credit watch on July 10, 2012, the ratings of 612 subprime RMBS with an original value of $7.35 billion, and two days later downgraded 498 of these securities.

On the same day, July 10, 2012, Moody's downgraded 399 subprime RMBS with an original value of $5.2 billion. Then, by the end of July 2012, S&P had downgraded more than 1,000 RMBS and almost 100 CDO securities. This is an unprecedented and huge volume of ratings. The downgrades created significant turmoil in the securitization markets because investors were required to sell off RMBS and CDO securities that had lost their investment grade status. In addition, the RMBS and CDO securities in the investment portfolios of financial firms lost much of their value while new securitizations were unable to find investors.

The subprime RMBS secondary market initially froze and then collapsed, as a result leaving financial firms around the world holding suddenly unmarketable subprime RMBS securities that were plummeting in value.[29]

In April 2007, both Moody's and S&P gave AAA ratings to three tranches (refers to the different types of mortgage-backed securities and CDO bonds that provide specified priorities and amounts of returns) of approximately $1.5 billion in a cash CDO known as Vertical Asset-Backed Securities (ABS) CDO 2007-1.[30]

Six months later, the report *Wall Street and the Financial Crisis: Anatomy of a Financial Collapse*, in its findings, stated that the majority of the CDO's tranches were downgraded to junk status; in 2008, the CDO's ratings were withdrawn, its assets were liquidated, and the AAA-rated securities became worthless.[31]

The fourth out of a total set of four identified causes of the financial crisis is listed as "Investment Bank Abuses: Case Study of Goldman Sachs and Deutsche Bank," and its appraisal is included in the report *Wall Street and the Financial Crisis: Anatomy of a Financial Collapse*. A significant and principal combination of events that influenced and energized the financial crisis was the role played and designed by complex financial instruments that are frequently identified and referred to as structured finance products, namely, (1) residential mortgage-backed securities (RMBS), or debt instrument secured by a pool of mortgages, whether residential or commercial, (2) collateralized debt obligations (CDOs), or the type of security often composed of the riskier portions of mortgage-backed securities, and (3) credit default swaps (CDS), or a type of credit derivative allowing a purchaser of the swap to transfer loan default risk to a seller of the swap; the seller agrees to pay the purchaser if a default event occurs, the purchaser does not need to own the loan covered by the swap.

The credit default swap contracts are linked to the ABX Index, which is a series of derivatives indexes constructed from the prices of twenty credit default swaps that each reference individual subprime mortgage-backed securities akin to an index like the Dow Jones Industrial Average. These financial products were envisioned, engineered, sold, and traded by major U.S. investment banks, and played a central role in the financial crisis.[32]

The report *Wall Street and he Financial Crisis: Anatomy of a Financial Collapse*, and its in-depth assessment, emphasized forcibly that two case studies, with particular reference to Goldman Sachs and Deutsche Bank, demonstrate a variety of troubling and sometimes abusive practices that involve the origination or use of RMBS, CDO, CDS, and ABX Indexes financial instruments.

Those practices, the report *Wall Street and the Financial Crisis: Anatomy of a Financial Collapse* affirmed, included at times the construction and the constructing RMBS or CDOs with assets that senior employees within the investment banks knew were of poor quality; underwriting securitizations for lenders known within the industry for issuing high-risk, poor-quality mortgages or RMBS securities; selling RMBS or CDO securities without full disclosure of the investment bank's own adverse interests; and causing investors to whom they sold the securities to incur substantial losses.[33]

Goldman Sachs, for example, used CDS and ABX contracts to place billions of dollars of bets that specific RMBS securities, baskets of RMBS securities, or collections of assets in CDOs would fall in value while at the same time, convincing customers to invest in new RMBS and CDO securities, thus, the unreasonable restraint of trade.

The credit default swap (CDS), a key factor in the expansion of the financial crisis, has skyrocketed in importance as a current structured finance product. Some investment banks modified an important structured finance product, which is defined as a derivative known as a credit default swap (CDS), for use in the mortgage market. A CDS, which is similar to an insurance contract, by definition is a contract between two parties in which one party guarantees payment to the other if the assets referenced in the contract lose value or experience a negative credit event.[34] The essence of the CDS product for the mortgage market was that it offered an alternative to investing in RMBS and CDO securities that would perform.

Another strategic alternative was brought forth. A consortium of investment banks under the leadership of both Goldman Sachs and Deutsche Bank launched in January 2006 a structured finance product that was connected to a newly created "ABX Index" that was designed to enable investors to bet on multiple subprime RMBS securities at once, according to the report *Wall Street and the Financial Crisis: Anatomy of a Financial Collapse*.[35]

The ABX Index, which increased risk, was administered by a private company called The Market Group and it included five separate indexes, each of which tracked the performance of a different basket of twenty designated subprime RMBS securities. For example, one index tracked a basket of twenty AAA- rated RMBS securities; the second index tracked a basket of AA-rated RMBS securities; and the remaining indexes tracked baskets of RMBS securities rated A, BBB, and BBB-. Every six months, a new set of RMBS securities was selected for each index.

The values of the securities in each basket, the report *Wall Street and the Financial Crisis: Anatomy of a Financial Collapse* found, were aggregated into a single composite value that rose and fell over time. Investors, for a fee, could take either the "long" position by betting on the rise of the index or the "short" position by betting on the fall of the index, without having to physically purchase or hold any of the referenced securities or raise the capital needed to pay for the full face value of those referenced securities.

The ABX Index magnified the risk by allowing investors to place unlimited bets on the performance of one or more of the subprime RMBS baskets. It also made it easier and cheaper for investors, including some investment banks, to short the subprime mortgage market in bulk.[36]

Investment banks not only helped establish the ABX Index, the investment banks used the Index themselves to bet on the mortgage market as a whole. In addition, the ABX Index expanded the risk inherent in the subprime mortgage market by providing investors with a way to make unlimited investments in RMBS securities.

THE COLLATERALIZED DEBT OBLIGATIONS (CDO) MACHINE: DEUTSCHE BANK, GOLDMAN SACHS, AND CITIGROUP

The role of the main parties in the current frame of reference of the collateralized debt obligations machine involves Deutsche Bank, Goldman Sachs, and Citigroup. The CDO machine impact did not cause the financial crisis; its effect was to make the financial crisis even worse.

The CDO built the transformation of the mortgage market through the process of creating a brand-new demand source for lower-rated tranches (or different types of mortgage-backed securities and CDO bonds that provide specified priorities and amounts of returns) of mortgage-backed securities.

CDOs are, by definition, structured financial instruments that purchase and pool financial assets such as the riskier tranches of various mortgage-backed securities; the CDO manager and securities firms pool various assets in an attempt to get diversification benefits, and just like mortgage-backed securities, the CDO issues the securities that vary based on their place in the cash flow bundle.

Deutsche Bank played an important role and was a major party in the CDO market and in the management of the CDO machine. One of the CDOs created by Deutsche Bank was called Gemstone CDO VII Ltd. (Gemstone 7) which provided the foundation and scale and scope of a specific CDO that added to understanding the CDO machine. Gemstone 7 issued its securities in March 2007 as a CDO; it was sponsored by HBK Capital Management (HBK), which was a large hedge fund that acted as the collateral manager for the CDO.

Gemstone 7 consisted of hybrid CDO with underlying variables that referenced a variety of high-risk, subprime residential mortgaged-backed securities (RMBS) initially valued at $1.1 billion hybrid CDOs when issued.[37]An estimated 90 percent of its assets were the mid- and subprime RMBS securities that included 33 percent carrying non-investment grade ratings. Of the remaining assets about 4.5 percent were CDO securities; 3.3 percent were commercial mortgage-backed securities; and 3.5 percent were securities backed by pools of student loans. In March 2007, Gemstone 7 had about $476 million in cash RMBS assets as well as $625 million in synthetic CDO assets.

Gemstone 7 was constructed in a way that some of its assets were set and could not change while other assets could be replaced by the collateral manager HBK, which is a Dallas-based hedge fund that was formed in October 1991 and managed about $12 billion in capital in 2007.[38]

HBK retained Deutsche Bank for the purpose of acting as the placement agent for Gemstone 7 in October 2006. The report *Wall Street and the Financial Crisis: Anatomy of a Financial Collapse* stated that under the agreement, HBK was to receive thirty basis points, or 0.3 percent (or $3.3 million), per year of the notional amount of Gemstone 7 in return for serving as the Collateral Manager

Most importantly, Deutsche Bank was to receive $6.79 million in underwriting fees but ultimately received a lesser amount of $4.7 million, according to the March 15, 2007, offering circular for Gemstone CDO VII Ltd. GEM 7, 00000427-816. "The Management Agreement at 168," the report *Wall Street and the Financial Crisis: Anatomy of Financial Collapse* provided documentation that, in fact, as opposed to an upfront fee, which Deutsche Bank reportedly received, the fee to HBK was reported to be paid quarterly. Therefore 0.3 percent of $1.1 billion translates to a fee of about $3.3 million for HBK.[39]

The cover page that concerns preference shares and identities Cede & Co. includes the following specific information:

- Form of Regulation S Global Share Certificate
- GEMSTONE CDO VII LTD.[40]
- Number Preference ISIN
- ( ) Shares ( )
- ( )
- Incorporated Under the Laws of the Cayman Islands
- US $845.00 Capital divided into 250 ordinary shares of a nominal par
- value of U.S. $1.00 per share and 59,500 preference shares of a par
- value of US $0.01 per share
- THIS IS TO CERTIFY THAT CEDE & CO. is the Registered Holder of 127
- Preference Shares in the above-named Company on this March 15, 2007.
- Executed on behalf of the said Company by:
- DIRECTOR ________________________________
- 40713-00080 NY: 19794343
- CONFIDENTIAL TREATMENT REQUESTED
- GEM 7-00001116

The inclusion of the statement, "Cede & Co. is the Registered Holder of Preference Shares," demanded the investigation that involved the significance of Cede & Co. published in the *Voting Rights in Major Corporations: A Staff Study* prepared by the Subcommittee on Reports, Accounting, and Management of the Committee on Governmental Affairs, U.S. Senate, 1978.[41] In fact, much of the stock on which voting rights are reported in the Staff Study is actually held by the Depository Trust Company in New York in the name of Cede & Co. (the name derives from Certificate Depository) was established by the financial community in the late sixties to simplify settlement of securities transactions among brokers and institutional customers. The Depository Trust Company (DTC) accepts stock from participating institutional depositors and then credits it to their accounts. Stock can be transferred within the DTC system by computerized book entry.

The cumbersome and costly movement of certificates thus became unnecessary. At the end of 1976 the Depository Trust participants included 40 banks and 211 broker-dealers. They held 3.8 billion shares of equity securities. However, DTC has no voting or discretionary rights whatsoever regarding this stock. The Depository Trust Company (Cede & Co.) does not vote stock it holds.

Depository Trust Co. (Cede & Co.) provides to corporations whose stock it holds "Security Position Listings," which identify each bank, brokerage firm, or other depository participant for whom DTC holds stock in the portfolio corporation. The listing shows the number of shares in the portfolio company held for each depository participant.

Deutsche Bank had many departments and desks that were in the management, designing, marketing, and selling the Gemstone 7 securities. The Deutsche Bank Collateralized Debt Obligations (CDO) Group, which was under the leadership of both Michael Lamont and Michael Herzig, directed the designing of its structure, monitoring of the purchasing and warehousing of its assets, obtaining its credit ratings, preparing the legal documentation, establishing its administrative structure, obtaining underwriting approval of the deal, designing the marketing materials, and overseeing the issuance of the CDO securities.[42]

The construction of CDO components relied on (a) securities firm, (b) CDO managers, (c) rating agencies, (d) investors, and (e) financial guarantors. The Report of the Financial Crisis Inquiry Commission (FCIC), pursuant to Public Law 111-21, January 2011, established that securities firms were underwriters of CDOs and approved the selection of collateral, structured the notes into tranches, and then sold them to investors.[43]

The underwriters placed emphasis on the generation of fees and structuring deals that they could sell while they took the risk that the asset might lose value. For example, Michael Lamont of Deutsche Bank's CDO Desk has remarked that the underwriter's business was to make new issue fees and to hedge if the market did have a downturn. It was not his position or responsibility to make decisions as to whether the rating agencies' models had the correct underlying assumptions.[44]

The CDO Manager's responsibility was to select the collateral such as mortgage-backed securities, and obtain periodic fees that were based on the dollar amounts of assets. The rating agencies' responsibility included providing basic guidelines on the collateral and the structure of the CDOs such as the sizes and returns of the various tranches while working closely with underwriters.

The role of the CDO investor triggered harmful consequences during the financial crisis when initially the CDO investors focused on different tranches based on their preferences for risk and return like triple-A (AAA) tranches. However, other CDOs became most important because CDO underwriters undertook the selling of tranches that were rated A, and especially, those rated-BBB tranches of mortgage-backed securities, the lowest and riskiest investment grade rating, thereby pushing up market prices on those BBB mortgage-backed securities.[45]

With the objective of issuing CDO securities, Deutsche Bank established an offshore corporation in the Cayman Islands called Gemstone CDO VII Ltd. To provide the management of the corporation, Deutsche Bank appointed its Cayman Island affiliate, Deutsche Bank Cayman, which is a licensed trust company. In its capacity as administrator, Deutsche Bank Cayman provide Gemstone 7 with the administrative services required to operate the CDO securitization, including, but not limited to, providing office facilities and secretarial staff, maintaining books and records required by Cayman law, naming at least two Cayman directors, and acting as the share registrar for Gemstone shares.[46]

Deutsche Bank International Limited is a wholly owned subsidiary of Deutsche Bank AG. It currently has several offices around the world, including one in the Cayman Islands, Deutsche Bank (Cayman) Limited ("DB Cayman"), that was opened in 1983.

One particular procedure gained in significance with regard to Gemstone 7 asset selection process. Most importantly, Deutsche Bank and HBK entered a private agreement in which Deutsche Bank agreed to provide adviser services to the issuer (Gemstone 7) and the company (HBK) on the selection and acquisition of the underlying assets. In addition, Deutsche Bank under both the engagement agreement and the separate risk sharing agreement also had the right to reject assets selected by HBK for the Gemstone warehouse account.[47]

Several problems began to occur in 2007 in connection with the mortgage industry. For example, the subprime lender Fremont Investment & Loan reported on January 29, 2007, that it severed ties with eight thousand brokers whose loans had high delinquency rates. One week later the subprime lender New Century announced on February 7, 2007, that its nonprime loan production would be declining and that New Century would be restating its earnings.[48]

As Deutsche Bank and HBK struggled to sell Gemstone 7 and were concerned about their exposure should Gemstone 7 not be fully subscribed, there was a plunge in the ABX Index, which fell from a high of $0.90 early in the month to a low of $0.69 by the end of the month that represented a decline of more than 23 percent in the value of subprime residential mortgage-backed securities (RMBS).

In March 2007 the mortgage market continued to be worse off even though Deutsche Bank's trading desk was one of the biggest traders in RMBS and CDOs on Wall Street; simultaneously, Deutsche Bank continued to market the Gemstone 7 securities.

Gemstone 7 closed on March 15, 2007. New Century Financial Corp., which had been a major provider of loans to people with risky credit, reported that it had lost support from its financial backers and was being delisted from the New York Stock Exchange (NYSE): New Century comprised 15 percent of Gemstone 7; in the meantime $400 million of Gemstone 7 was unsold.[49]

Gemstone 7 received credit ratings that were produced by Standard & Poor's (S&P) and Moody's on the same day it closed, March 15, 2007. In the following Table Zeta 06: Table of Gemstone VII Ratings by Tranche, the top three tranches, which is a representation of 73 percent of the value of the Collateralized Debt Obligations (CDOs), in fact, received triple-A (AAA) ratings.[50]

Table 5.1. Table of GEMSTONE VII Ratings by Tranche

Tranche	*Initial Ratings: Date*	*1st Down-grade: Date*	*2nd Down-grade: Date*	*3rd Down-grade: Date*
	AAA	A+	BB+	CC
Class A-1a	Mar. 15, 2007	Feb.5, 2008	July 11, 2008	Aug.19, 2009
	AAA	B–	CC	
Class A-1b	Mar. 15, 2007	Feb.5, 2008	July 11, 2008	N/A
	AAA	AA–	CCC–	CC
Class A-2	Mar. 15, 2007	Nov. 21, 2007	Feb. 5, 2008	July 11, 2008
	AA	BBB	CC	
Class B	Mar. 15, 2007	Nov. 21, 2007	Feb. 5, 2008	N/A
	A	B–	CC	
Class C	Mar. 15, 2007	Nov. 21, 2007	Feb. 5, 2008	N/A
	BBB	CCC+	CC	
Class D	Mar. 15, 2007	Nov. 21, 2007	Feb. 5, 2008	N/A
Class E	Mar. 15, 2007	Nov. 21, 2007	Feb. 5, 2008	N/A

- Preference Not rated
- Shares
- Source: S&P

In the Initial Rating column, the next three tranches received investment grade ratings of AA, A, and BBB despite the fact that one-third of the CDOs' underlying assets carried non-investment grade ratings.

An examination of the "1st Downgrade" column in Table Zeta 06: Table of Gemstone VII Ratings by Tranche shows a pattern of Gemstone 7 losses as five of its seven tranches were downgraded in November 2007, which is eight months after Gemstone VII closed on March 15, 2007, and the downgrade included one of its triple-A (AAA) tranches.

The pattern in the "2nd Downgrade" column of Table Zeta 06: Table of Gemstone VII Ratings by Tranche showed that by July 2008 all seven tranches had been downgraded to junk status, thereby leaving the Gemstone 7 securities just about worthless.[51]

Deutsche Bank issued fifteen new CDOs with total assets of $11.5 billion during the period from December 2006 through December 2007. Gemstone 7 was only one of many CDOs that Deutsche Bank assembled and was the underwriter for as the mortgage market deteriorated in 2007. In effect, Deutsche Bank was the fourth-largest issuer of CDOs in the United States.

Goldman Sachs multiplied the damaging effects of the collapse in subprime mortgage lenders, and together with other major U.S. investment banks' abuses and predatory practices, in fact, they powered their position of dominance through the utilization, origination, trading and selling, and underwriting of what are called complex financial instruments or structured finance products.

For example, the residential mortgage-backed securities (RMBS), collateralized debt obligations (CDOs), synthetic collateralized debt obligations, and credit default swaps (CDS), along with the CDS contracts that relinked to the ABX Index, are assembled, sold, and traded under "structured finance products." Major investment banks established the ABX Index that permitted counterparties to wager on the rise or fall in the value of a basket of subprime RMBS securities and that could be used as a frame of reference for the condition of the subprime mortgage market as a whole.

Goldman Sachs became connected with a number of various troubling, abusive, and unreasonable restraint of trade practices that were generated out of its involvement in the origination, underwriting, selling and trading, or unreasonable restraint of trade utilization of RMBS securities, CDOs, CDS, and ABX Index financial instruments.[52]

Those troubling and abusive practices, the report Wall Street and the Financial Crisis: Anatomy of a Financial Collapse stated, were driven by constructing RMBS or CDOs with assets that were described by senior employees in the investment bank as being of poor quality; underwriting securitizations for lenders who are known within the investment bank industry for using high-risk, poor-quality mortgages or RMBS securities; selling RMBS or CDO securities without full disclosure of the investment bank's own adverse interests; and causing investors to whom they sold the securities to incur substantial losses.[53]

With particular reference to Goldman Sachs, those troubling and abusive practices included exploitation and exploiting conflicts of interest with Goldman Sachs's own clients. For instance, Goldman Sachs used CDS and ABX Index of dollars of bets that specific RMBS securities, baskets of RMBS securities, or collections of assets in CDOs would fall in value, while at the same time, Goldman Sachs was convincing its customers to invest in new RMBS and CDO securities.[54]

Under the direction of both Goldman Sachs and Deutsche Bank, a consortium of investment banks started a different structured finance product in January 2006 that was connected with a brand-new "ABX Index" for the purpose of enabling investors to bet on multiple subprime RMBS securities at once.

The ABX Index was managed in 2006 and 2007 by a private company known as the Market Group Ltd. that included five separate indexes, each of which tracked the operational performance of a different basket of twenty designated subprime RMBS securities. One index tracked a basket of twenty triple-A–rated RMBS; the second index tracked a basket of double-A rated RMBS securities and the remaining indexes tracked baskets of RMBS securities rated A, BBB, and BBB-. Every six months, a new set of RMBS securities was selected for each index.

The values of the securities in each basket were aggregated into a single composite value that rose and fell over time. For compensation, investors could take either the "long" position, betting on the rise of the index, or take the "short" position, betting on the fall of the index, without having to physically purchase or hold any of the referenced securities or raise the capital needed to pay for the face value of these referenced securities. The ABX Index allowed investors to place unlimited bets on the performance of one or more of the subprime RMBS baskets. Goldman Sachs possessed the total short position, or the sale of a borrowed security with the expectation that the asset will fall on value, thus of a $2 billion CDO called Hudson-1, then selected assets for the CDO to transfer risk from Goldman Sachs' own holdings, permitted investors to purchase the CDO securities while never completely disclosing Goldman Sachs own short position, as a result and when the CDO lost value, Goldman Sachs made a $1.7 billion gain at the expense of the clients to whom Goldman had sold the securities.

In a separate instance, Goldman Sachs traded and sold a synthetic CDO called ABACUS 2007-AC-1 to clients never disclosing that the firm had allowed the sole short party in the CDO, which was a hedge fund, to play a dominant part in the selection of the assets. As expected by Goldman Sachs, the ABACUS securities speedily lost value and the hedge fund made an estimated $1 billion in profit while the three long investors together lost $1 billion.

Goldman Sachs provides managerial services to sophisticated institutional investors, financial services companies, pension funds, hedge funds as well as a limited number of very wealthy individuals.

Looking back from a historical perspective, Goldman Sachs which was established in 1869 as an investment bank and private partnership, became a publicly traded corporation, and expanded its operations after the enactment of the Gramm-Leach-Bliley Act of 1999 that repealed the Glass-Steagall Act and the latter's restrictions placed on the activities that investment banks could be engaged.

It was in 2008 when Goldman Sachs took advantage of the Gramm-Leach-Bliley Act (1999) provisions and converted to a Bank Holding Company. In 2007 Goldman Sachs reported net revenues in the amount of $11.6 billion; $3.7 billion of this net revenue figure was generated by the Structured Products Group in the Mortgage Department, driven largely as a result of its subprime investment activities.

Over the last decade, Goldman Sachs has placed its emphasis on being a Wall Street trading house that provides broker-dealer services to institutional customers, acting as a prime broker to hedge funds, structuring and financing deals for customers from its own capital, and conducting proprietary trading activities for its own benefit.[55]

Henry Paulson, the former Chief Executive Officer of Goldman Sachs before he became the Secretary of the U.S. Treasury Department in 2006, addressed subprime mortgages with the characterization that many bad loans had been issued before his appointment as the Secretary of Treasury.

Furthermore, Henry Paulson continued, subprime mortgages went from accounting for 5 percent of total mortgages in 1994 to 20 percent by 2006. Securitization, or the process of pooling debt assets such as mortgages, auto loans, and credit card debt into a separate legal entity that then issues a new financial instrument or security for sale to investors, had separated originators from the risk of the products that they originated. The result of this finding, Henry Paulson noted, was a housing bubble that eventually burst in far more spectacular fashion than most previous bubbles.[56] While Henry Paulson was CEO at Goldman Sachs objectives centered on the origination and sale of mortgages securities. Billions of dollars in loans were provided to mortgage lenders from 2004 through 2006; most of the billions of dollars in loans were allocated to the subprime lenders among whom are Ameriquest, Long Beach, Freemont, New Century, and Countrywide through the procedure of warehouse lines of credit, frequently in the form of repurchase agreements (Repos), which means a method of secured lending where the borrower sells securities to the lender as collateral and agrees to repurchase them at higher price within a short period, often within one day.[57]

The period 2004 through 2006 was also the time that Goldman Sachs acquired $53 billion in the form of loans from the preceding list of subprime lenders as well as other subprime loan originators, which Goldman Sachs securitized, or pooling of assets into securities backed by those assets, and sold to investors.

Simultaneously, Goldman Sachs issued 318 mortgage securitizations that amounted to $184 billion, an estimated one-fourth included subprime mortgages, and 63 Collateralized Debt Obligations (CDOs) registering at $32 billion. In the continuation of this pattern between 2004 and 2006, Goldman Sachs issued 22 synthetic or hybrid CDOs with a face value of $35 billion.[58]

The unreasonable restraint of trade practices that Goldman Sachs and the Synthetic Collateralized Debt Obligations (CDOs) "marriage" produced the suppression of competition activity trough the control of the noncompetitive outcome; thus, the nullification of competition according to the author's assessment.

What are synthetic CDOS?—the response to this question was provided in the Financial Crisis Inquiry Report, *Final Report of the National Commission on the Causes of the Financial and Economic Crisis in the United States* (Official Government Edition), January 2011, which defined synthetic CDOs as complex paper transactions involving credit default swaps. They are different from the regular or traditional cash CDO. The synthetic CDOs contained no actual tranches of mortgage-backed securities, or even tranches of other CDOs.[59]

Instead, synthetic CDOs referenced these mortgage securities and therefore the synthetic CDOs were bets on whether borrowers would pay their mortgages. In the place of real mortgage assets, these synthetic CDOs contained credit default swaps (CDS) and did not finance a single home purchase. The credit default swaps are defined as a type of derivative allowing a purchaser of the swap to transfer loan default risk to a seller of the swap. The seller agrees to pay the purchaser if a default event occurs. The purchaser does not need to own the loan covered by the swap.

The investors in these synthetic CDOs included "funded" long investors, who paid cash to purchase actual securities issue by the CDO; "unfunded" long investors, who entered into swaps with the synthetic CDO and making money if the reference securities performed; and "short" investors, who bought credit default swaps on the reference securities and making money if the securities failed. Funded investors received interest if the reference securities performed; they could lose all of their investment if the reference securities defaulted.[60]

The Goldman Sachs and Synthetic CDO "marriage" produced its first major synthetic CDO in 2004, it was called Abacus 2004-1 and was representative of a deal that was valued at $2 billion. An estimated one-third of the swaps referenced residential mortgage-backed securities, about another third referenced the existing CDOs and the remainder consisted of commercial mortgage-backed securities (along with bundled commercial real estate loans) and other securities.

Goldman Sachs was the short investor (short selling means to sell a borrowed security in te expectation of a decline in value). For the entire $2 billion deal, Goldman Sachs purchased credit default swap protection on these referenced securities from the CDO. There were funded investors such as IKB (a German bank), the TCW Group, and Wachovia Bank that put up a total of $195 million to purchase Mezzanine Tranches of the deal.

These investors, IKB (a German bank), TCW Group and Wachovia would receive scheduled principal and interest payments if the referenced assets performed. Alternatively, if the referenced assets did not perform, then, Goldman Sachs, as the short investor, would receive the $195 million. Within this frame of reference, the IKB, TCW Group and Wachovia were "long" investors and betting that the referenced assets would perform well, and Goldman Sachs was a "short" investor and betting that they would fail.[61]

In effect, the long investors in Abacus 2004-1 anticipated the receipt of millions of dollars if the referenced securities performed. Alternatively, Goldman Sachs expected to gain an estimated $2 billion if the referenced assets failed. The result is that Goldman Sachs, the short investor in the Abacus 2004-1 CDO, received an estimated amount of $930 million. The long investors lost nearly all of their investments.[62]

In April 2008, Goldman Sachs received $7.3 million as payment from GSC as a result of CDS protection sold by GSC to Goldman on the first and second loss tranches. In June 2009,Goldman Sachs was paid $806 million by AIG Financial Products as a result of the credit default swap (CDS) protection it had purchased against the Super-Senior Tranche. TCW paid Goldman Sachs $23 million as a result of the CDS bought against the Junior Mezzanine Tranches; Goldman Sachs received $30 million from IKB because of the CDS it bought against the C Tranche; IKB paid Goldman an additional $40 million in April 2010 as a result of the CDS against the B Tranche.[63]

During the period that covered July 1, 2004, and May 31, 2007, Goldman Sachs packaged and sold forty-seven synthetic CDOs with an aggregate face value of $66 billion, according to the Report Submitted by the Financial Crisis Inquiry Commission, Pursuant to Public Law 111-21, January 2011. Shorting many of these deals is the way Goldman Sachs earned profits; in other deals, Goldman Sachs would earn profits by facilitating the transaction between the buyer and seller of credit default swap protection.

There is evidence that has been uncovered and shows the detailed transactions leading to Goldman Sachs' short positions that were undertaken to advance only the proprietary financial interests of Goldman Sachs and not as a function of its market making role to assist clients in buying or selling assets.[64] The failure of many of its RMBS and CDO securities that were underwritten and sold, moreover, served s the source of profits for Goldman Sachs because of shorts. Other than the Abacus 2007-AC1 collateralized debt obligation, Goldman Sachs CDO activities included (a) Hudson Mezzanine 2006-1, (b) Anderson Mezzanine 2007-1, and (c) Timberwolf 1.

Hudson Mezzanine 2006-1 (Hudson 1) was a $2 billion synthetic CDO comprised of $1.2 billion in ABX (a series of derivatives indexes constructed from the prices of twenty credit default swaps that each reference individual subprime mortgage-backed securities) assets from the inventory of Goldman Sachs, and $800 million in single-name CDS contracts on subprime RMBS and CDO securities that Goldman wanted to short.

By definition, a mezzanine CDO means the referenced RMBS securities carried the riskier credit ratings of BBB or BBB-. In this instance, Goldman Sachs used the CDO to transfer the risk associated with its ABX assets to the investors that purchased Hudson-1 securities. At the same time, 100 percent of the short side of the CDO was taken by Goldman Sachs, an indication, thereby, that Goldman Sachs would profit if any of the Hudson-1 securities lost value.

Most significantly, Goldman Sachs exercised complete control over the CDO especially in its own establishment and administration of the CDO and in its capacity as the underwriter, initial purchaser of the issued securities, senior swap counterparty, credit protection buyer, collateral put provider, and liquidation agent.[65] From its inception, Hudson-1 had been designed for the purpose of removing substantial risk from the balance sheet of Goldman Sachs; in addition, it took 100 percent of the short position in the CDO, an investment which is contrary to the interest of investors in Hudson 1 securities.

The major investor in Hudson-1 was Morgan Stanley which lost an estimated $960 million in Hudson-1 during 2008 as a result of Goldman Sachs delay for months in selling the Hudson 1 assets even though the Hudson-1 assets had lost value and received credit rating agencies downgrades. The short position of Goldman Sachs increased in value while the assets of Hudson-1 declined in value.[66]

The over-the-counter (OTC) derivatives market was unregulated and substantially opaque; there were no public reporting requirements and barely any price discovery was operational. The OTC derivatives market participants grew weary and concerned involving the exposures and trustworthiness of their counterparties and the value of their respective contracts. The trading and selling in the OTC derivatives markets experienced a downturn in response to investors becoming disturbed and concerned in connection with counterparty risk while hedge funds and other participants in the market experienced reducing their positions or exited the market. The market activity decreased to the level of a crawl. There was not any market at all in some cases—no trades whatsoever. Thus, a sharp and unprecedented contraction of the market occurred.[67]

Trading on the OTC derivatives market came to a grinding halt; it shutdown resulting in jeopardizing the viability of every participant irrespective of their direct exposure to subprime mortgage-backed securities. Moreover, the participants, in effect, reacted by liquidating their positions in other assets that those swaps were designed to hedge.

Severely damaged results hampered the OTC derivatives market that had been driven by nonprime mortgages. Previously, companies had placed substantial reliance on the prices of OTC derivatives contracts that were included in the ABX Indexes in order to value their nonprime mortgage assets.

In the fall of 2008 nearly every firm to reduce exposure to every other firm because the contraction of the OTC derivatives market produced implications beyond the valuation of mortgage securities. In addition, the OTC derivatives had been applied in the utilization and management of all manner of risk such as the risk that currency exchange rates could fluctuate, the risk that interest rates would change, and the risk that asset prices would rise.

The OTC derivatives had served the purpose of efficiently managing these various types of risk which required liquidity so that positions could be adjusted daily at little cost.[68]

Between December 2007 and June 2010 the credit derivatives shrunk by 48 percent, from their height of $58.2 trillion in notional amount to the low of $30.4 trillion, respectively.[69]

The global derivatives market is huge, unregulated since 2000, and at the end of June 2008, the global OTC derivatives market registered at $673 trillion in notional amount while the gross market value amounted to $20 trillion.[70]

The determination, availability and measurement of risk in the OTC derivatives market were not disseminated and distributed to market participants or government regulators such as the Federal Reserve Bank, thereby, forming the enlargement of the scope of uncertainty and lack of essential OTC derivatives information, based on the Financial Crisis Inquiry Commission Report Final Report of the National Commission on the Causes of the Financial and Economic Crisis in the United States. The demand has increased involving the search for information connected with the interlocking directorates that OTC derivatives created among the small number of large firms that act as dealers in the OTC derivatives business. The major OTC derivatives trading volume positions and the resulting counterparty (a party to a contract) credit and operational risks, were concentrated in a few large firms as of June 30, 2008. For instance among the U.S. bank holding companies the following institutions held enormous OTC derivatives positions, (a) J. P. Morgan $94.5 trillion in notional amount, (b) Bank of America $37.7 trillion, (c) Citigroup $35.8 trillion, (d) Wachovia $4.1 trillion, and (e) HSBC $3.9 trillion.

Shortly thereafter, and in the first quarter of 2009 and after having become bank holding companies officially in 2008, Goldman Sachs held $45.9 trillion and Morgan Stanley held $37.0 trillion in notional amount of OTC derivatives.

This Financial Crisis Inquiry Commission Report, Pursuant to Public Law 111-21, January 2011, has drawn attention to the then 2008 current and potential exposure to OTC derivatives, and especially credit default swaps, at the top five United States bank holding companies that were on average three times greater than the capital that the five top bank holding companies had on hand to meet the regulatory requirements. This Financial Crisis Inquiry Commission Report noted in its findings that the risk was even higher at the investment bank. For example, Goldman Sachs had OTC derivatives exposure which was more than 10 times its capital.[71]

The concentration of financial, monetary and economic power dominance positions under control of the major bank holding companies and investment banks, in fact, posed risks, increased uncertainty and threats of unreasonable restraint of trade practices for the financial and monetary and banking system, and because of their likelihood and further expansion of mergers, acquisitions, holding companies, and interlocking directorates with other financial institutions engaged in the elimination of competition involving the trading and selling OTC derivatives and credit default swaps.

MARKET STRUCTURE, CONCENTRATION OF CONTROL, INTERLOCKING DIRECTORATES AMONG MAJOR U.S. CORPORATIONS AND FINANCIAL CRISIS

The ramifications from Wall Street and *The Financial Crisis: Anatomy of a Financial Collapse*, and the Report by the Financial Crisis Inquiry Commission have portrayed representations and subsets of the market structure of corporate ownership and concentration among the nation's largest corporation, in effect, identified by the major institutional shareholders and interlocking directorates in one hundred of the largest corporations, and described by the extent that such ownership and interlocking directorates provide substantially the potential for a high degree of corporate concentration and global competitive advantage. The market structure is the embodiment and encompasses the industrial organization economics characteristics that are the source that determined the vertical integration, horizontal integration, and conglomerate that exist between sellers to sellers, buyers to buyers, sellers to buyers, and the overt and tacit pricing behavior conduct between sellers and new entrants in the market. The market structure dimensions, most importantly, address the degree of seller concentration, or the number one, few or many sellers, the same for buyers, and the condition of entry or the height of barriers to new entrants.

In this section and stage of the book, the author heightens the importance of industrial organization economics sis in the formulation and implementation of a frame-of-reference, platform-complete strategy of the market structure, concentration of ownership and control, and intrlocking directorates in the U.S. economy and the financial, economic, and monetary crisis.

1980 STRUCTURE OF CORPORATE CONCENTRATION INSTITUTIONAL SHAREHOLDERS AND INTERLOCKING DIRECTORATES AMONG MAJOR U.S. CORPORATIONS

Among the numerous sources that are available for examination, assessment, and utilization in the author's principal designing characteristics of the book include the 96th Congress, 2d Session, *Structure of Corporate Concentration, Institutional Shareholders and Interlocking Directorates Among Major U.S. Corporations: A Staff Study*, Committee of Governmental Affairs, U.S. Senate, Volume 1, December 1980, Washington, DC: U.S. Government Printing Office, 1980.

This 1980 Staff Study covers two general areas of corporate concentration: institutional shareholders and interlocking directorates; it is the potential control over one company or over many companies that is the major focus of this 1980 Staff Study; the results identify the top institutional investors in each of the major stock corporations under analysis, and identify for each of the one hundred companies in the Staff Study all of their direct interlocks, and those indirect interlocks that it has with institutional investors and each of the other companies in this Staff Study.

INSTITUTIONAL CONCENTRATION

This 1980 Staff Study of the structure of corporate concentration makes a documentation contribution and provides valuable insights through the following examples of institutional concentration. At the top of the structure of American Industry, J. P. Morgan is the most dominant investor and, in 1979, held $18.5 billion worth of equities in its portfolio.

A critical finding of this Staff Study brought forth that J. P. Morgan held an even more powerful position of institutional dominance among the stock companies; it was the number one institutional investor in 15 companies. It placed 43 times among the top six shareholders and 72 times among the top 25 holders. There was no other institution that came close in terms of potential equity power.[72] Moreover, the 1980 Staff Study has affirmed that J. P. Morgan in the area of commercial banking was the number one shareholder in Bank of America, which was the nation's largest holding company; J. P. Morgan was the number one shareholder in Citicorp, the nation's second largest banker, and the number three equity holder in Manufacturers Hanover, the fourth largest U.S. banking company.

J. P. Morgan directors interlocked with the directors of these three "competitive" banks in the following manner; Bank America, 11 times on 8 intermediate boards; Citicorp, 35 times on 20 boards; and Manufacturers Hanover, 23 times on 14 boards.

J. P. Morgan, among the energy companies, was the largest shareholder in Mobil; the number three shareholder in Exxon; the number four shareholder in Gulf Oil; and the number six shareholder in Standard of California; J. P. Morgan directly interlocked with Mobil.

Citicorp was the number one shareholder in Chase Manhattan (33 indirect interlocks). Furthermore, Citicorp was the fourth largest shareholder in both Bank America and Manufacturers Hanover.

Among the energy companies, Citicorp was the number two shareholder in both Standard of Indiana and Atlantic Richfield, and, in fact, Citicorp directly interlocked with Exxon, Mobil, and Standard of California.[73]

Among the retailing companies, J. P. Morgan was the top institutional shareholder in Mobil, the owner of Montgomery Ward, the top shareholder in J. C. Penney, and number one shareholder in Kmart.

In the airlines industry, J. P. Morgan was the top institutional shareholder in both Eastern Airlines and Pan American. In the aerospace and aircraft, J. P. Morgan was the leading stockholder in United Technologies, the number three shareholder in Rockwell International, the number five institutional shareholder in Lockheed, and the number six shareholder in Boeing. Under the industry leaders classification, J. P. Morgan was the number one institutional shareholder in Citicorp, General Electric, IBM, and United Technologies; J. P. Morgan was the number two shareholder in General Motors, number three stockholder in Exxon, and the number two stockholder in Sears.[74]

This 1980 Staff Study has made a contribution of valuable insights as well as heightened comprehension involving corporate concentration. For example, the major companies in the Staff Study provided evidence that they were found to be connected not only through common major shareholders but also by extensive interlocking directorates. The number one company by size in an industry was also the dominant one in sales, assets, net income, and employees, over its largest competitor. In addition, the dominant company was about double the size of its next competitor as well as three to four times larger than the other major competitors.[75]

In ranking the companies by size of the major one hundred companies, the largest companies were found to have more top institutional shareholders in common with their competitors along with a greater number of competitor interlocks on third-party boards of directors, thereby, in the author's assessment, resulting in an unreasonable restraint of trade.

This 1980 Staff Study delivered an assessment of corporate concentration with particular emphasis on New York banking companies. Citicorp is the largest New York banker; it is two-thirds the size of Chase Manhattan, which is its nearest competitor, and more than twice the size of the next three financial giants, namely, Manufacturers Hanover, J. P. Morgan, and Chemical New York, while it is three times the size of Bankers Trust.[76]

In regard to the concentration through common ownership and the interlocking directorates of the New York banks, the Staff Study results showed that Citicorp was the number one institutional investor in Chase Manhattan. The directors of Citicorp were interlocked with directors of Chase Manhattan 33 times in the boardrooms of twenty companies that included powerful customers such as AT&T, Exxon, General Motors, General Electric, Sears and IBM. Citicorp was the number one investor in Bankers Trust with linkages 21 times on 10 boards of directors, which included AT&T, IBM, Mobil, J. C. Penney, and Xerox.

Furthermore, Citicorp was the number four stockholder in Manufacturers Hanover, and interlocked with the latter company 26 times on 16 boards of directors. Citicorp was the number five shareholder in Chemical New York with 27 interlocks on 15 boards. Citicorp was the number seven shareholder in J. P. Morgan, number four in BankAmerica, number three in Continental Illinois, and number two in Harris Bankcorp.

Among the New York banking companies, more importantly, the top stockholders in Citicorp were J. P. Morgan and Prudential, which together accounted for over 5 percent of Citicorp's stock; thus, J.P. Morgan interlocked with Citicorp 35 times on 20 boards of directors that included General Motors, General Electric, IBM, and Mobil, as well as DuPont and U.S. Steel. Prudential connected with Citicorp 13 times on 9 boards, including AT&T, Exxon, General Electric, and Xerox. J. P. Morgan was also the third largest stockholder in Manufacturers Hanover with whom Morgan indirectly interlocked 23 times. J. P. Morgan was the number one stockholder in BankAmerica.[77]

The findings of this 1980 Structure of Corporate Concentration Staff Study have added recognition to the fact that the top shareholders in J. P. Morgan were National Detroit, T. Rowe Price, First National Bank of Boston, Chase Manhattan, BankAmerica, Continental Corp., and Citicorp, which represented a combined power of 16 percent of the outstanding shares. Citicorp and Chase Manhattan together controlled nearly 4 percent of the stock of Morgan while National Detroit was interlocked with J. P. Morgan 3 times on the board of General Motors. The Chase Manhattan Bank interlocked with J. P. Morgan 19 times on 11 boards of directors that included General Motors, General Electric, Metropolitan Life, and IBM. BankAmerica interlocked with Morgan 11 times on 8 boards of directors.

COMPANY DIRECT AND INDIRECT INTERLOCKS AMONG NEW YORK BANKING COMPANIES

This 1980 volume, *Structure of Corporate Concentration, Institutional Shareholders and Interlocking Directorates Among Major U.S. Corporations: A Staff Study*, provides a complete listing of direct and indirect interlocks among one hundred companies under study. In the enumeration of the links between two corporations, the Staff Study affirmed all direct interlocks are listed first, followed by any indirect links. A direct interlock occurs between two firms whenever a board member of one company sits on the board of the other company.

An indirect interlock exists whenever a board member of a company sits on the board of another company together with a director who is also a board member of a third corporation. The first firm is then indirectly linked to the third company through the second or intermediate company, the company on whose board both directors sit.

The company direct and indirect interlocks among the largest of the New York banking companies had boards of directors with broad representation from potential financial customers in the major U.S. industries. This 1980 Structure of Corporate Concentration—Institutional Shareholders Among

Major U. S, Corporations A Staff Study has underscored that sitting together on the board of directors, for example, of Citicorp were directors from three of the nation's largest nuclear energy companies, General Electric, Westinghouse, and Stone & Webster; they were joined by a director of National Nuclear Ltd., and General Electric, Ltd.[78]

The interlocking directorate relationship, which is a representation of a second area of corporate concentration, enabled the advancement actions for the establishment of intercompany communications as well as consensus. The linkages at the boardroom meetings, moreover, demonstrated personal interconnections through which essential information would be passed; thus, arrangements would be completed and policies formed. Among the major energy corporations, Exxon was represented on the Citicorp board by its chairman. Mobil was represented on the Citicorp board along with Halliburton (2), American Natural Resources, and Texas Eastern.

The largest of the New York banking companies had boards of directors in the transportation sector that were directors from General Motors (3), Union Pacific (2), Southern Pacific, Boeing (2), and United Technologies (4). The chemical industry was represented by the chairman of DuPont, the president of W. R. Grace, and two directors from Monsanto. In the basic metals industry there were directors from U.S. Steel, Aluminum Company of America (Alcoa), Kennecott Copper (2), and Phelps Dodge.[79]

On the Citicorp Board of Directors, the American Telephone & Telegraph Company (AT&T) had three directors, and one director was its former chairman. In addition to AT&T, there was its leading competitor, IBM, with two directors who were joined by directors from Xerox (2), Sperry (2), NCR, Kodak (2), and Exxon. Large competitors were represented on the board of directors of Morgan Bank . For instance, in insurance there were directors situated on the Morgan Bank Board of Directors from Prudential, Metropolitan Life, New York Life, John Hancock, n INA. Separately, General Motors and Ford were represented on the Morgan Bank Board of Directors.

In nuclear energy, the company interlocks among the largest of the New York banking companies were Bechtel, General Electric, and Union Carbide. On the Morgan Board of Directors were U.S. Steel and Bethlehem Steel in addition to Aluminum Company of America, Kennecott Copper, and Goodyear.

From the perspective of shareholder concentration and within the scope of examples of concentration, the top six institutional shareholders in each of the automobile companies held substantial concentrations of equity power, such as in General Motors where they totaled 5.3 percent of the outstanding shares; in the Ford Motor Company they aggregated an estimated 8 percent (without the Ford family interest); the top six institutional shareholders held over 12 percent in Caterpillar, and the top six institutional shareholders totaled about 8 percent in International Harvester.

Common ownership was dominant in shareholder concentration among the automotive companies. For example, TIAA-CREF was the number one institutional shareholder in General Motors, the number three stockholder in Ford, and the number three stockholder in Caterpillar. Common ownership in shareholder concentration among automotive companies included J. P. Morgan as the number two shareholder in General Motors, the top shareholder in Caterpillar, and shareholder number four in Ford and number nine in International Harvester. The third-largest institutional shareholder in General Motors was National Detroit, which was also the number five shareholder in Caterpillar. Capital Group was the number six shareholder in General Motors, the second-largest investor in Ford, and the number three shareholder in International Harvester.

These specific forms of common ownership positions held by the top six institutional shareholders in each of the automotive companies were strengthened through the utilization of interlocking directorates relationships between the major institutional shareholders and their portfolio companies.[80]

More particularly, TIAA-CREF was the number one investor in General Motors and, simultaneously, was directly interlocked with General Motor's second-largest institutional shareholder, which was J. P. Morgan. In addition to and joining with TIAA-CREF on the board of directors of J. P. Morgan were directors of General Motors (3), Ford, and Prudential, the fifth- largest shareholder in General Motors. TIAA-CREF had an interlocking directorate connection with Citicorp, the second-largest stockholder in Chrysler. Also on the board of directors of Citicorp were three directors of General Motors. TIAA-CREF, the number one shareholder in General Motors, directly interlocked with Chase Manhattan, the fifth-largest shareholder in Chrysler.

Seated on the Chase Manhattan Board of Directors were directors of General Motors and Chrysler. National Detroit, which was the number three shareholder in General Motors, the number nine in Chrysler, and the number five in Caterpillar, was directly interlocked with General Motors and Chrysler (twice). The largest shareholder in J. P. Morgan was National Detroit, and Morgan was the number two shareholder in General Motors. DuPont interlocked with General Motors on the boards of directors of Morgan, Citicorp, AT&T, and Bethlehem Steel.[81]

The top six institutional shareholders in each of the automotive companies also held ownership positions in the suppliers to the industry. For example, Capital Group was the sixth- largest institutional investor in General Motors, the second largest in Ford Motors, and the third largest in International Harvester. In steel, the top institutional investors in the automotive companies that held ownership positions in suppliers to the automotive industry included Capital Group as the lead shareholder in U.S. Steel, Bethlehem Steel, and Armco Steel; moreover, TIAA-CREF was the number two shareholder in U.S. Steel and the number six stockholder in Bethlehem Steel, and Morgan was the number one in LTV (which owned Jones & Laughlin Steel).

Directors of U.S. Steel and Bethlehem Steel held seats on the board of directors of Morgan (along with General Motors and Ford). Both U.S. Steel and Bethlehem Steel were directly interlocked on the board of directors of General Motors.

In the context of the rubber industry, the Capital Group was the number one shareholder in Goodyear and the number two shareholder in Firestone. TIAA-CREF was the second-largest shareholder in Goodyear and the third largest shareholder in Firestone. A director of Goodyear held a seat on the board of directors of Morgan, as did directors representing General Motors, Ford, U.S. Steel, and Bethlehem Steel.

Among the top institutional investors in the electric equipment industry, Morgan was the number one shareholder and TIAA-CREF was the number two shareholder in General Electric, while Capital Group was ranked first and Morgan was second in Westinghouse Electric.[82]

The interlocking directorates in the automotive companies demonstrated connections with each other through extensive interlocks on the boards of directors of major suppliers, customers, and financial institutions, in addition to the sharing of major investors. For example, three General Motors directors sat with a Ford director on the board of directors of J. P. Morgan and on its bank board, Morgan Guaranty Trust. In addition, General Motors interlocked with Chrysler on the boards of directors of Chase Manhattan, the Royal Bank of Canada, National Detroit, and the New York Stock Exchange.

General Motors connected with both Chrysler and Caterpillar on the board of directors of AT&T; General Motors linked with Chrysler and International Harvester and Caterpillar on the board of Santa Fe Industries. Ford interlocked with Chrysler and International Harvester on the board of Hewlett-Packard and with Chrysler alone on Kmart and Utah International. Thus, General Motors indirectly interlocked with Ford 10 times on 5 boards of directors; General Motors indirectly interlocked with Chrysler 13 times on 9 boards; General Motors indirectly interlocked with Caterpillar 15 times on 11 boards of directors and indirectly interlocked with International Harvester 7 times on 5 boards of directors.[83]

The automotive companies' interlocking directorates had substantial interlocks with other companies;, General Motors had 27 direct interlocks and 873 indirect interlocks with 99 companies, according to the 1980 Staff Study. Ford had 11 direct and 315 indirect interlocks, Chrysler had 14 direct and

386 indirect interlocks, Caterpillar had 17 direct and 401 indirect interlocks, and International Harvester had 23 direct and 317 indirect interlocks. In total, the five automotive companies directly interlocked 92 times and indirectly interlocked 2,352 times among the 100 companies selected for the 1980 Staff Study.

On the General Motors Board of Directors were seats held by the representatives of many of the largest banking companies in North America, such as Citicorp, Chase Manhattan, Morgan, National Detroit, and the Royal Bank of Canada. With respect to customers, there were four large customers who held seats on the board of directors of General Motors: AT&T, Canadian Pacific, Santa Fe Industries, and Union Pacific. There were three top airlines with seats on the General Motors Board of Directors, namely, American Airlines, Braniff International, and Northwest Airlines.

From the standpoint of telecommunications, there were seats on the board of directors of General Motors held by AT&T and Communications Satellite. Insurance companies, including Metropolitan Life and Sun Alliance, had directors on the General Motors board. With reference to the pharmaceutical industry, there were five pharmaceutical companies represented on the board of directors of General Motors: Allied Chemical, Ltd., Johnson & Johnson, Merck, Pfizer, and Warner-Lambert.[84]

Among the numerous examples of corporate concentration presented in the 1980 Staff Study, an appraisal is made of the energy industry and the nation's powerful seven largest energy companies in which Exxon held dominance over its six competitors. With its $50 billion in assets at the time of this 1980 Structure of Corporate Concentration Staff Study, Exxon was and Texaco; in addition, Exxon was an estimated three times the size of each of the next major companies (Gulf, Standard of Indiana, and Standard of California), and Exxon was about three and one-half times the size of Atlantic Richfield. Essentially, all seven of the energy companies were among the twelve largest industrials in the nation. The other five largest industrials were General Motors, Ford, IBM, General Electric, and ITT, the makeup of the top twelve category in assets.[85]

The shareholder concentration represented by the stockholding of the energy industry's six largest institutional investors were especially concentrated. For instance, Exxon's top six shareholders, namely, Chase Manhattan, Manufacturers Hanover, Morgan, TIAA-CREF, Fayez Sarofim, and Citicorp, had an aggregate investment power of about seven percent of its outstanding shares which amounted to a total value od $1.7 billion. The three largest Exxon shareholders accounted for more than $1 billion of that value and over four percent of the outstanding shares. Combined with this investor power, the six top owners of Exxon had a total linkage of three direct interlocks and at least thirty-seven indirect interlocks boardroom ties to their portfolio company.

Under the category of shareholder concentration, Mobil's top six investors managed a total of 8.5 percent of Mobil stock and had one direct and nineteen indirect interlocks with the company. Similarly, the concentrated stockholdings at the top of the remaining four energy competitors ownership sizes were as follows, Texaco's top six holdings had 4.9 percent; Standard of California's top six holdings had 15.9 percent; Gulf's top six holdings had 12.1 percent; and Standard of Indiana's top six holdings had 6.5 percent; also Atlantic Richfield's top six holdings had 10.6 percent.[86]

It is of immense importance to note the fact that, from the perspective of shareholder concentration, the largest stockholders in Exxon also held top equity positions among Exxon's competitors, thereby reflecting an unreasonable restraint of trade practice. More particularly, the 1980 *Structure of Corporate Concentration: Institutional Shareholders and Interlocking Directorates: A Staff Study* reported that Chase Manhattan Bank, which is the number one stockholder in Exxon, was the number four stockholder in Mobil, number two stockholder in Standard of California, and the number three stockholder in Standard of Indiana.

The members of the Chase Manhattan Bank Board of Directors held seats on the boards of directors of Exxon and Standard of Indiana, and were directly interlocked with the Rockefeller family interests. Exxon's number two shareholder was Manufacturers Hanover, which is also the third largest shareholder in Atlantic Richfield, the fourth largest in Standard of California, and the sixth largest shareholder in Texaco. Of great significance is that members of the board of directors of Manufacturers Hanover were represented on the boards of directors of Standard of Ohio, Cities Service, and Royal Dutch Petroleum.[87]

Shareholder concentration and market power dominance combination were strengthened as a result of J. P. Morgan as the third largest shareholder in Exxon; also J. P. Morgan was the number one shareholder in Mobil, number four shareholder in Gulf, number six shareholder in Standard of California, and number eight shareholder in Texaco.

TIAA-CREF was the number four shareholder in Exxon, number two shareholder in Texaco, and the number eight shareholder in Standard of California. Furthermore, TIAA-CREF directly interlocked with Exxon, Chase, and Morgan.

Exxon's number five shareholder was Fayez Sarofim, a large Houston-based investment company that was the largest stockholder in Standard of Indiana. Fayez Sarofim also was the number three shareholder in Standard of California; number four shareholder in Texaco; number five shareholder in Mobil; number six shareholder in Gulf; and number seven shareholder in Atlantic Richfield. Fayez Sarofim reported no director interlocks.

Citicorp was ranked as the sixth largest stockholder in Exxon, number two stockholder in Standard of Indiana, number two stockholder in Atlantic Richfield, and number nine stockholder in Texaco. In regard to interlocking directorates, Citicorp directly interlocked with Exxon and Mobil, and had two directors on the board of directors of Standard of California.

The 1980 Staff Study, Committee on Governmental Affairs, U.S. Senate, in its findings provided valuable insights on the importance of the electric equipment manufacturers, especially General Electric and Westinghouse Electric that were major factors in the energy industry, specifically in connection with nuclear energy. In fact, General Electric and Westinghouse Electric had shareholders in common with the oil companies (all of which also control coal and uranium resources). Most significantly, J. P. Morgan ranked as the number one shareholder in General Electric, and the number two stockholder in Westinghouse Electric.

In connection with interlocking directorates, J. P. Morgan directly interlocked with General Electric three times and had eight indirect interlocks with General Electric on five boards of directors. J. P. Morgan indirectly interlocked with Westinghouse Electric four times on three boards of directors.

The TIAA-CREF was the number two shareholder in General Electric. In addition, Citicorp was the number three shareholder in General Electric; National Detroit was the number five shareholder in General Electric. In total, the top six shareholders in General Electric represented an estimated 7 percent of the shares outstanding.

In regard to Westinghouse Electric, the Capital Group was the primary shareholder, and next in line were Morgan and Bank of America representing total shareholdings of more than 8 percent of the Westinghouse Electric stock. The top six shareholders accounted for 11.3 percent of the shares outstanding.

The energy companies' interlocks, which are evidence of corporate concentration, have deepened and expanded the importance of interlocking directorates through their dominance and control of institutional shareholder concentration. Based on the 1980 Staff Study, there existed significant concentration among the company competitors in the degree that the energy companies connected with each other on the boards of directors of the nation's largest corporations.

For example, the members of the board of directors of Exxon met with the Mobil directors on the boards of directors of AT&T, Caterpillar Tractor, Chemical Bank, Chemical New York, Citicorp, Citibank, Equitable Life, and Procter & Gamble, while Exxon participated in interlocking directorates and interlocked with Texaco on the boards of directors of the Arabian-American Oil Company and TIAA-CREF. Exxon connected with Standard of California and connected again with Standard of California on the board of directors of Arabian-American Oil Company, Caterpillar Tractor, and Citicorp. Exxon interlocked with Standard of Indiana on the boards of directors of Chase Manhattan, Chemical New York, and Continental Illinois. In addition, Exxon interlocked with Atlantic Richfield on the boards of directors of Caterpillar Tractor, Kraft, Metropolitan Life, PepsiCo, and Scott Paper.

Mobil, the nation's second-largest energy producer, indirectly interlocked with its major competitors thirty-three times. Mobil interlocked with Exxon on eight intermediate boards including the major companies such as AT&T, Caterpillar Tractor, Chemical New York, Citicorp, Equitable Life, and Procter & Gamble.

Mobil interlocks connected with General Electric eighteen times on ten corporate boards. Furthermore, Mobil connected with Standard of California on the boards of Boeing, Caterpillar Tractor, Citicorp, and Rand Corp.

Mobil linked with Texaco on the board of American Express and interlocked with Standard of Indiana on the board of Chemical New York, Dow Jones, and Northern Trust. Mobil interconnected with Gulf on the board of the Aluminum Company of America, and it interlocked with Atlantic Richfield on the board of directors of Caterpillar Tractor.[88]

The nation's seven largest energy companies, as determined by the 1980 Staff Study included Exxon, Mobil, Texaco, Gulf, Standard of Indiana, Standard of California, and Atlantic Richfield, also known as "the seven energy giants," had 64 direct interlocks and 2,318 indirect interlocks.

With the one hundred companies in the 1980 Staff Study. There existed the following groupings of the seven energy companies' boards of directors that were seated at the boards of directors of major financial institutions.[89]

1. BankAmerica: Standard of California, Atlantic Richfield, Union Oil and Getty
2. Citicorp: Exxon, Mobil, Standard of California, General Electric and Westinghouse
3. Chase Manhattan: Exxon, Standard of Indiana, General Electric and Westinghouse
4. Manufacturers Hanover: Cities Service, Standard of Ohio, General Electric and Westinghouse
5. J. P. Morgan: Mobil, Cities Service, Louisiana Land & Exploration, General Electric, Bechtel Group and Tenneco
6. Chemical New York: Exxon, Mobil, Standard of Indiana, Shell Oil, General Electric and Westinghouse

7. Bankers Trust: Mobil, Conoco, Newmont Mining and Hudson Bay Oil & Gas
8. Western Bancorporation: Standard of California, Atlantic Richfield and Caltex Petroleum
9. Continental Illinois: Exxon, Standard of Indiana and Conoco
10. Metropolitan Life: Exxon, Atlantic Richfield, Phillips Petroleum, Cities Service, and Standard Oil of Ohio.

The financial crisis of 2007 and 2008 have roots in and institutions driven by the 1980 Staff Study, and the scale and scope of the financial and economic crisis 2007 and 2008, for example, are embedded in the energy company boardrooms provided meetings places for some of the nation's largest financial institutions. Represented on the Exxon Board of Directors were Citicorp, Chase Manhattan, Chemical New York, Continental Illinois, Equitable Life, Metropolitan Life, and Prudential, among other institutions. Mobil's board of directors was host to Citicorp, Morgan, Chemical New York, Bankers Trust, and Equitable Life. Standard of California's board of directors included Bank of America, Citicorp, Crocker, and Western Bancorporation. Standard of Indiana hosted directors from Chase Manhattan, Chemical New York, Continental Illinois, First Chicago, Harris Bankcorp and Northern Trust.

The "telecommunications and information processing and office equipment industry" findings in the 1980 Staff Study, affirmed that the American Telephone and Telegraph Company (AT&T) towered far above all the companies in these fields, and AT&T was the world's largest private communications conglomerate. For example, AT&T accounted for more than 80 percent of all telephone service in the United States as of the 1980 Staff Study. AT&T owned the largest communications equipment manufacturer in the U.S. as it operated the nation's largest communications research laboratories, and employed one million workers.[90]

Of great significance is the emphasis placed on shareholder concentration in the "telecommunications and information processing and office equipment industry" that is deepened and strengthened because of the following identified shareholders dominance; the top shareholder in AT&T was TIAA-CREF, which was also the top voting shareholder in the three major competitors, such as General Telephone & Electronics (GTE), International Telephone and Telegraph (ITT), and Radio Corporation of America (RCA).

TIAA-CREF was the number six shareholder in IBM, which was the closest competitor of AT&T, and interlocked with Citicorp, Chase and Morgan. Shareholder concentration ascendency is underscored by the dominance of Morgan, who was the number one shareholder in the Radio Corporation of America (RCA), although Morgan ranked as number two in voting authority at RCA behind TIAA-CREF; Morgan was the top shareholder in International Business Machine (IBM) and was the fifth largest shareholder in International Telephone and Telegraph (ITT).

In the field of information equipment, Morgan was the number one stockholder in Sperry, number one stockholder in Xerox, number four stockholder in Honeywell, and number four stockholder in Eastman Kodak. Simultaneously, Morgan directly interlocked with IBM, Western Electric, a subsidiary of AT&T, and Eastman Kodak.

Other top shareholders in AT&T were among the largest of the telecommunications industry competitors and consisted of (a) Manufacturers Hanover Bank, the third-largest shareholder in AT&T, the number two shareholder in IBM, the number three shareholder in GTE, the number two shareholder in both Xerox and Eastman Kodak; (b) Citicorp was the fourth largest shareholder in AT&T, the fourth largest shareholder in IBM and the number one shareholder in Honeywell; (c) Bankers Trust was the fifth largest shareholder in AT&T and the number three shareholder in IBM; and (d) Chase Manhattan Bank was the fifth largest shareholder in IBM, number two shareholder in ITT, the number three shareholder in Sperry, and the number six shareholder in Eastman Kodak. In total, all four top shareholders in AT&T, namely, Citicorp, Chase Manhattan, Morgan, and Bankers Trust, had directors on the board of directors of IBM.

Exxon is a key competitor in the field of information processing and office equipment. In fact, Exxon had the following major shareholders-in-common with communications and information companies, (i) Morgan (IBM, ITT, RCA, SPERRY, Honeywell, Xerox, and Eastman Kodak); (ii) Citicorp (AT&T, IBM, Honeywell, and Eastman Kodak); (iii) Chase (IBM, ITT, RCA, Sperry, and Eastman Kodak); (iv) Manufacturers Hanover (AT&T, IBM, GTE, ITT, RCA, and Eastman Kodak).

From the perspective of Exxon as a key competitor in the information field there existed a director of Exxon who was on the board of directors of AT&T. The directors of Exxon joined the directors of both AT&T and IBM seated on the board of directors of Caterpillar Tractor, Chase Manhattan Bank, Chemical New York Bank, Citicorp, and Continental Illinois. Furthermore, Exxon connected with IBM, Eastman Kodak, Sperry, and RCA on the board of directors of Metropolitan Life.[91]

AT&T dominated the interlocking directorates and controlled the greatest number of interlocks with the 100 companies selected for the 1980 Staff Study. In essence, AT&T had 39 direct interlocks and 938 indirect interlocks with the 100 companies. There were 20 direct interlocks and 733 indirect interlocks by IBM with the 100 companies in the 1980 Staff Study; Eastman Kodak had 15 direct interlocks and 162 indirect interlocks; Xerox had 14 direct interlocks and 379 indirect interlocks with the 100 companies.[92] The largest number of interlocking directorates as well as the most interconnections and information processing and office equipment, thus, have derived from the seating of multiple competitors on the boards of directors of the largest customers, suppliers, and financial institutions. In addition, the members of the boards of directors of AT&T, IBM, Exxon, Xerox, Sperry, and Eastman Kodak were seated on the board of directors of Citicorp.

The directors of IBM, RCA, Sperry, Eastman Kodak, and Exxon were represented on the board of directors of Metropolitan Life. The directors of IBM, AT&T, Xerox, and GTE were seated on the board of directors of Bankers Trust. The members of the board of directors of AT&T, IBM, Eastman Kodak, and Exxon were represented on the board of directors of Continental Illinois Bank. The directors of AT&T, IBM, RCA, and Exxon sat together on the board of directors of Chemical New York.

The members of the board of directors of AT&T, IBM, and GTE were represented on the board of directors of Conoco. Most significantly, the dominant competitor interlocking directorates conduct and performance were between AT&T and IBM with 30 interlocks on 16 intermediate boards of directors.[93]

There existed a high degree of concentration demonstrated on AT&T's own board of directors with directors who represented AT&T's major customers, suppliers, and financial institutions. For example, the major banks with representation on the board of directors of AT&T were Bankers Trust, Chase Manhattan, Chemical New York, Continental Illinois Bank, Crocker National, First City Bancorp, Manufacturers Hanover Bank, U.S. Trust, Wachovia Bank, and Western Bancorporation. Furthermore, the industrial companies with representation on the board of directors of AT&T were Exxon, Mobil, Conoco, Tenneco, General Motors, Chrysler, Caterpillar Tractor, U.S. Steel, Inland Steel, DuPont, Lockheed, Eastman Kodak, Union Carbide, and Corning Glass.

In addition, the insurance companies represented on the board of directors of AT&T were Prudential, John Hancock, American Express, Continental Corp., and Pacific Mutual. The retailers on the board of directors of AT&T were represented by J. C. Penney, Federated Department Stores, Carter Hawley Stores, Levi Strauss, and Hart Schaffner & Marx.

VOTING RIGHTS IN MAJOR CORPORATIONS AND THE FINANCIAL CRISIS 2007 AND 2008

The voting rights of 122 major corporations that included their investment decision making, in fact, have strengthened concentration of control, supported advancement in industrial and economic growth, formed the base and a source for corporations' expansion, formation, and transnational management of major corporations and multinational enterprises involving global competitive advantage.

In this key publication, 95th Congress, 2d Session, U.S. Senate, Document No. 95-99, *Voting Rights in Major Corporations: A Staff Study*, prepared by the Subcommittee on Reports, Accounting, and Management of the Committee on Governmental Affairs, January 1978, Washington, DC: U.S. Government Printing Office, April 24, 1978, there is a cornerstone based on and inherent in its accompanying industrial organization economics analysis, the concentration of control of voting rights-to-stock in large U.S. corporations, and the voting rights dominance of the large institutional investors.

This 1978 Staff Study has ranked the identified voters in 122 corporations according to their voting rights, which include the largest financial, industrial, transportation, utility, and retailing firms in the nation, and some of the largest insurance companies and investment complexes.

The market value of the common stock of the 122 corporations amounted to 41 percent of the market value of all outstanding common stock in the

Table 5.2a. Extent of Stockvoter Interests of 21 Major Investors in the 122 Corporations Studied

Name of Investor	*Number of Companies in Which Among Top 5*	*Number of Times Investor is Stockvoter No. 1*
Morgan Guaranty Trust Co. of New York	56	27
Citibank	25	7
TIAA/CREF (Teachers Insurance & Annuity Association-College Retirement Equities Fund)	24	2
Capital research & Management Co.	19	2
Prudential Insurance Co. of America	18	4
Dreyfus Corp.	17	4
National Bank of Detroit	17	5
Kirby Family Group-Allegheny Corp.	16	4
BankAmerica Corp.	15	1
Fidelity Management & Research Corp.	13	2
Manufacturers Hanover Trust Co.	12	1
Bankers Trust Co.	11	0
First National Bank of Chicago	11	2
Lord Abbett & Co.	11	2

Table 5.2b. Extent of Stockvoter Interests of 21 Major Investors in the 122 Corporations Studied

Name of Investor	*Number of Companies in Which Among Top 5*	*Number of Times Investor is Stockvoter No. 1*
Equitable Life Assurance Society of the United States	10	2
First National Bank of Boston	10	0
Harris Trust & Saving Bank	10	2
Chase Manhattan Corp.	8	3
Continental Illinois National Bank & Trust	8	3
Marlennan Corp. (Marsh & McLennan Companies, Inc.)	7	0
Massachusetts Financial Services Inc.	7	2

United States at the end of 1976, which is indicative of the enormous size of the 122 corporations.[94]

Table Eta 07: Extent of Stockvoter Interests of 21 Major Investors in the Total of 122 Corporations Studied[95] is a summary or frame of reference that portrays the concentration and significance of voting rights included in the Staff Study.

This source of the preceding Table Eta 07: Extent of Stockvoter Interests of 21 Major Investors in the 122 Corporations Studied is the 1978 Staff Study prepared by the Subcommittee on Reports, Accounting, and Management of the Committee on Governmental Affairs, U.S. Senate, Document Number 95-99, April 24, 1978, and referenced above.

The Morgan Guaranty Trust Co. of New York's dominant position soars above all others in the extent of stockvoter interest of 21 major investors in the 122 corporations studied in the identified 1978 Staff Study. In the above Table Eta 07, Morgan Guaranty Trust Company of New York is included in the top five identified stockvoters in 56 or nearly half of the 122 corporations. Morgan is stockvoter number one in almost one-fourth or 27 of the 122 corporations.

After Morgan is Citibank, which ranked among the top five stockvoters in 25 corporations and ranked as the major identified stockvoter in seven corporations. The twenty-one institutions listed in the above Table Eta 07 that dominated the ranks of investors include eleven banks, five investment company complexes, four insurance companies, and the family group which controlled the world's largest investment company complex.

The 1978 Staff Study stated the question, "who controls the major voting rights in the dominant banks?" and provided the answer, "the principal stockvoters in large banks are large banks." The Morgan Guaranty Trust Co. of New York is stockvoter number one in four of its New York sister banks, including Citicorp, Manufacturers Hanover Corp., Chemical New York Corp., and Bankers Trust New York Corp. Moreover, Morgan Guaranty is stockvoter number one in Bank of America Corp., which is the bank holding company with the largest assets. Most significantly, Morgan Guaranty's average voting rights in each of these respective banks exceed 3 percent.

Alternatively, Citicorp is stockvoter number one in Morgan Guaranty's parent holding company, J. P. Morgan & Co. The Chase Manhattan Bank is stockvoter number two in J. P. Morgan & Co.; Chase Manhattan is stockvoter number three and number four in J. P. Morgan & Co. and Manufacturers Hanover, and Bankers Trust in whose parent bank holding companies Morgan Guaranty Trust is stockvoter number one.[96]

The Depository Trust Company (DTC) in New York, in the name of Cede & Co., holds much of the stock on which voting rights were reported in this 1978 Staff Study. The financial community created the Depository Trust Company (DTC) in the late 1960s for the purpose of the simplification of the settlement of securities transactions that involved brokers and institutional customers. The Depository Trust Company (DTC), in the management of its operations, accepts stock from the participating institutional depositors and then credits it to their accounts. Inside the DTC system the stock could be transferred through computer processing and computerized book entry.

The Depository Trust Company (DTC) delivers "Security Position Listing" for corporations whose stock the DTC holds; the Security Position Listing identifies each bank, brokerage firm or other depository participant for whom the DTC holds stock in the portfolio company held for each depository participant. The number of participants that were included in the Depository Trust consisted of 40 banks and 211 broker dealers that, together, held 3.8 billion shares of equity securities at the end of 1976. In spite of the shares of equity securities held, the DTC had no voting or discretionary rights at all regarding the stock held.

CONCENTRATION OF CONTROL THROUGH INTERLOCKING DIRECTORATES AND THE FINANCIAL CRISIS

There are several sources of industrial organization economics that have made a contribution to the assessment of the concentration of economic, financial, political and market power and control in the United States. The author, within this context, offers the following additional documentation for showing the extent of concentration of control of corporate power among the largest corporations and among the major companies in certain industries trough emphasis placed on the institution of interlocking directorates.

The source of the additional documentation for showing the extent of concentration of control of corporate power among the largest corporations and among the major companies in certain industries consists of 95th Congress, 2d Session, U.S. Senate, Document No. 95-107, *Interlocking Directorates Among the Major U.S. Corporations: A Staff Study*, prepared by the Subcommittee on Reports, Accounting, and Management of the Committee on Governmental Affairs, June 15, 1978, Washington, DC: U.S. Government Printing Office, 1978. There are several reasons given that have been enumerated as the necessity for the completion of the above identified 1978 Staff Study and among which are included (a) investigations have identified the interlocking directorate as a corporate device with the potential for harm in the United States competitive economy, (b) corporations which are linked together directly, or indirectly, enjoy special channels of communications that approach the elimination of competition, preferential treatment in the buying and selling of goods and financial services followed by the concentration of undue economic power among the few, (c) the corporate management interlocks create problems of conflicts of interest, in this connection, a director who serves on the board of directors of different corporations experience the divided loyalties; a director's opportunity for inside dealings could become an act of subordination of the interests of stockholders for his private benefit; a director of two companies that have business dealing with each other, or whose interests otherwise may be in conflict with each other, is placed in the anomalous position of being a fiduciary with respect to each other, and (d) there is a problem that involves the deterioration in the performance by the directors who are seated on the boards of directors of too many corporations, especially the larger corporations. They may not have sufficient time and the preparation for required decision making.

This 1978 Staff Study, along with its analysis, involves only direct and indirect interlocks among companies through their directors; it does not cover interlocks that involve officers who are not directors, nor does it reach down to directors of subsidiaries or affiliates of the companies; [97] the placing of emphasis in this referenced 1978 Staff Study heightened the importance of the directorate concentration among the 130 largest U.S. corporations.

There were 130 companies chosen for this 1978 Staff Study, selected from among the largest corporations included in *Fortune Magazine*'s rating by assets for the year 1976, and included the following classifications: industrials; commercial banks; life insurance companies; diversified-financial companies;

utilities; transportation companies; retailing organizations; and investment advisors. The 130 companies consisted of 115 of the 122 corporations that were examined in the 1978 Voting Rights Staff Study.[98]

For the purpose of this 1978 Interlocking Directorates Staff Study, the definition of interlock occurs when a director of one company is also a director of another company. Therefore, the two companies have a common director and through that director there is the opportunity for communication and understanding between the companies.

An indirect interlock occurs when the two companies, each has a director on the board of directors of a third—or intermediate—company, or conversely, where the third company has a director on the boards of each of the two companies. As a result and in such situations, there is also the opportunity for direct communication and shared knowledge between the first and second company, even though they are indirectly, rather than directly, interlocked.

It is important to note that this 1978 Interlocking Directorates Staff Study does not make any allegations as to the predatory use of specific interlocks for such anticompetitive purposes. There are certain general patterns of corporate concentration described and some potential for abuse identified in this 1978 Interlocking Directorates Staff Study's relationship to persons who were directors of the major companies during the 1976 calendar year. Some persons who were directors may not have held such positions during the entire year; other persons may not have continued their directorships beyond 1976.

The financial institutions classification is singled out for assessment at this particular stage and as foundations connected with financial institutions that are central to the financial institutions operations during the financial crisis 2007 and 2008. The 130 corporations that comprised the 1978 Interlocking Directorates Staff Study experienced a detailed and specific examination of the direct and indirect interlocks that involved the leading financial organizations such as, the 20 largest commercial banking companies, the 10 largest diversified financial companies, the 10 largest life insurance companies, and the 7 largest investment advisory firms.

The financial institutions classification scale and scope, in this instance, involved 16 limited major financial institutions that have been chosen from the overall 130 corporations listed in this 1978 Interlocking Directorates on the basis of their national importance, size, and potential for providing financial and investment services and capital to the country's largest enterprises.

One significant finding, as a result of studying the financial institutions classification, affirmed that the commercial bankers were not seated on each other's boards of directors. Instead, the commercial bankers met on the boards of directors of large nonbanking companies such as insurance companies, industrials, utilities, and transportation companies, thereby, making the creation of indirect interlocking directorate relationships gain the appearance of the delivery of ready-made channels for communication, business agreements, knowledge producing, and information, and thus, reciprocity (reciprocal dealing) which could lead to barriers to entry, pooling agreements, rigged bidding, tying arrangements, and coordinated rather than free entry demanded competitiveness financial policies among the nation's largest banking institutions.

For example, Citicorp, the nation's largest international bank holding company, connected with, as of 1978, its closest rival, the Chase Manhattan Bank, through 18 direct routes on 12 intermediate boards of directors. Most importantly, Citicorp interlocked with J. P. Morgan (which is the parent of the nation's largest institutional investor, Morgan Guaranty Trust Co.) 39 times on 19 other boards of directors. Simultaneously, Citicorp indirectly connected with Manufacturers Hanover 26 times on 15 boards of directors; Citicorp indirectly connected with Chemical New York 24 times on 16 boards of directors and Citicorp connected with Bankers Trust 14 times on 9 boards of directors.[99]

The Chase Manhattan Bank possessed a pattern of 18 indirect interlocks with Citicorp. In addition, the directors of Chase Manhattan met with directors who represented J. P. Morgan 17 times on 9 boards of directors. Similarly, Chase Manhattan interlocked with Manufacturers Hanover 27 times on 13 boards of directors. Chase Manhattan interlocked with Chemical New York 16 times on 12 boards of directors. Chase Manhattan Bank interlocked with Bankers Trust 7 times, interlocked with First Chicago 10 times and interlocked with Continental Illinois Bank 6 times.

J. P. Morgan, in addition to its substantial indirect links with Citicorp and Chase Manhattan Bank, indirectly interlocked with Manufacturers Hanover 25 time on 15 boards of directors; J. P. Morgan interlocked with Chemical New York 15 times on 9 boards of directors and interlocked with Bankers Trust 6 times; J. P. Morgan interlocked with First Chicago 9 times.

Table 5.3. The 16 Major Financial Institutions–Direct and Indirect Interlocks among Competitors Identification Make-up this List[1]

Bank America	J.P. Morgan
Bankers Trust	Western Bancorporation
Chase Manhattan Bank	Aetna Life & Casualty
Chemical	Equitable Life
Citicorp	John Hancock
Continental Illinois	Metropolitan Life
First Chicago	New York Life
Manufacturers Hanover	Prudential

[1] Ibid., p. 122

The most indirect interlocks between the New York banking companies, and reported by the 1978 Interlocking Directorates Staff Study, were those between Manufacturers Hanover and Chemical New York that included 32 routes on 17 intermediate boards of directors. The Manufacturers Hanover also indirectly connected with Bankers Trust 11 times, and indirectly connected with First Chicago 8 times.

The Continental Illinois and First Chicago Corp. banking companies have directors who have met together on 18 other boards of directors for a total of 42 indirect interlocks which represented the most extensive interconnection between two financial companies found in the 1978 Interlocking Directorates A Staff Study; they are not directly interlocked.

The Continental Illinois Bank also directly interlocked with AT&T on whose board of directors was seated the directors from Citicorp, J. P. Morgan, Manufacturers Hanover, Chemical New York, and Prudential, along with some other financial companies. The First Chicago had a direct interlock with General Motors, on whose board of directors were seated the directors of Citicorp, Chase Manhattan, J. P. Morgan, and Metropolitan Life, among others.[100]

Within the total of the six largest insurance companies that are included under the overall financial institutions classification, there were no direct interlocks among the six largest insurance companies, however, opportunities for their linkages on other company boards of directors were very substantial. For example, the directors of the Metropolitan Life Insurance board met the representatives of the New York Life Insurance Company Board of Directors 17 times on 12 intermediate boards of directors. Metropolitan Life indirectly interlocked with John Hancock 11 times on intermediate boards of directors and indirectly interlocked with Equitable Life 9 times. Metropolitan Life indirectly interlocked with Prudential 6 times and indirectly interlocked with Aetna Life & Casualty 8 times, all on the intermediate boards of directors. Prudential, which is the largest among the six largest insurance companies, indirectly interlocked with Equitable Life 5 times, and indirectly interlocked with New York Life 5 times.

The interlocking directorates relationship between the insurance companies and the bankers demonstrated a more intensive strengthening of conduct and interorganizational behavior pattern. For example, Metropolitan Life, which was the second-largest life insurance company, was directly represented on the board of directors of Citicorp, Chase Manhattan, J. P. Morgan, and twice on the board of Chemical New York—the second-, third-, fifth-, and sixth-largest bank holding companies, ranked by assets respectively, in the nation. Furthermore, Metropolitan Life was indirectly interlocked with each of these banking companies an average of 17 times. Metropolitan Life possessed some other direct interlocks with First Chicago, Charter New York, Crocker

National, Mellon National, Fidelity Union Bancorporation, First National Holding Corp., First Wisconsin Corp., Lincoln First Banks, and the Royal Bank of Canada. The New York Life Insurance Co. was more prolific than Metropolitan Life. For example, New York Life Insurance Co. had three directors on the board of directors of Manufacturers Hanover, and two directors seated on the Chemical New York Board of Directors. In essence, the boards of directors of Citicorp and J. P. Morgan each had a director represented from New York Life Insurance Co. When emphasis is centered on the indirectly interlocked relationships, then, the New York Life Insurance Co. indirectly interlocked 21 times with Manufacturers Hanover; indirectly interlocked 18 times with Chemical New York; indirectly Interlocked 9 times with Chase Manhattan; and also indirectly interlocked 5 times with Bank of America.[101]

The Prudential, which is the largest life insurance company, directly interlocked with J. P. Morgan and Bankers Trust. However, Prudential indirectly interlocked with Citicorp 14 times and indirectly interlocked with Chase Manhattan Bank 10 times; Prudential had 9 indirect routes to Chemical Bank. A key insurance company–bank interlocking relationship involved John Hancock Mutual Life with First National Boston Corp. The 1978 Staff Study in its findings reported that 5 direct interlocks (including the chairman of the board of directors of John Hancock) and 27 indirect interlocks, in fact, bound together John Hancock Mutual Life with First National Boston Corp. Furthermore, on the First National Boston Corp. Board of Directors, the John Hancock Mutual Life interfaced with five directors of Liberty Mutual Insurance Co.; with two directors of New England Mutual Life; with one director of Massachusetts Life; and with directors of four other insurance companies. On the First National Boston Corp. Board of Directors were seated the representatives of three major utilities, and the chairman of a utility construction firm, Stone & Webster.

The strengthening of the degree of concentration of control increased in scale and scope when the portrayal involved the following financial interlocks on major nonfinancial boards and banks represented on insurance company boards of directors.[102] The following is a listing of some of these corporations together with the names of the major financial companies represented on their boards of directors.

The 1978 *Interlocking Directorates Among the Major U.S.Corporations: A Staff Study* addressed the "Financial Interlocks on Major Nonfinancial Boards" by as follows: American Telephone &Telegraph and Western Electric (100%-owned subsidiary of AT&T): Citicorp, Manufacturers Hanover (2), Chemical New York (2), Bankers Trust (2), Western Bancorporation, Prudential Insurance, John Hancock Mutual Life, American Express, and Continental Corp. General Motors: Citicorp (2), Chase Manhattan (2), J. P.

Morgan (3), First Chicago, Mellon National, National Detroit (2), Metropolitan Life (2), Northwestern Mutual. Ford: Citicorp, J. P. Morgan, Bankers Trust, Wells Fargo, Equitable Life, and John Hancock Mutual Life. Chrysler: Bank of America, Chase Manhattan (2), Manufacturers Hanover (2), National Detroit (2), Continental Corp., and New York Life. United States Steel: Citicorp, Chase Manhattan, Chemical New York, Manufacturers Hanover, Mellon National, Charter New York, First Bank System, Metropolitan Life, Aetna Life & Casualty (2), and American Express. Bethlehem Steel: Citicorp, J. P. Morgan, Continental Illinois, and Metropolitan Life. DuPont: Citicorp, Bank of America, J. P. Morgan, Chemical New York, and John Hancock Mutual Life. General Electric: Citicorp, Chase Manhattan, J. P. Morgan (3), Manufacturers Hanover, Chemical New York, Security Pacific, Wells Fargo, Prudential, and American Express. Westinghouse Electric: Citicorp, Manufacturers Hanover, Chemical New York, Mellon National, Aetna Life & Casualty, and New York Life. Exxon: Citicorp, Chase Manhattan, Chemical New York, Aetna Life & Casualty, Equitable Life Assurance, Metropolitan Life, and Prudential Insurance. Mobil: Citicorp, Chemical New York, Bankers Trust, Equitable Life, and American Express. Atlantic Richfield: Chemical New York, First Chicago, Security Pacific, Metropolitan Life, and New York Life. Standard Oil Co. (Indiana): Chase Manhattan, Chemical New York, First Chicago, and Continental Illinois. Union Carbide: Chase Manhattan, J. P. Morgan, Manufacturers Hanover (2), Metropolitan Life (2), New York Life, and Continental Illinois Corp. (2). Sears Roebuck: Citicorp, Chase Manhattan, J. P. Morgan, Manufacturers Hanover (2), Chemical New York, Continental Illinois, First Chicago (2), and Crocker National. Southern Pacific: Citicorp, Manufacturers Hanover, Wells Fargo, Western Bancorporation, and New York Life (3). Burlington Northern: J. P. Morgan, First Chicago, First Bank System, Northwest Bancorporation, Equitable Life Assurance, and New York Life (2). International Business Machine: Citicorp (2), Chase Manhattan, J. P. Morgan, Chemical New York (2), Bankers Trust, Continental Illinois, Metropolitan Life, asnd American Express. Eastman Kodak: Citicorp, J. P. Morgan, Continental Illinois, Marine Midland, Metropolitan Life, and Aetna Life & Casualty.

The 1978 Staff Study addressed "Banks Represented on Insurance Company Boards" as follows: Metropolitan Life: Citicorp, Chase Manhattan, J. P. Morgan, Chemical New York, Charter New York, Crocker National, First Chicago, and Mellon National. Equitable Life Assurance: Citicorp, Chase Manhattan, and Mellon National. Prudential Insurance: J. P. Morgan, and Bankers Trust. Aetna Life & Casualty: Citicorp and J. P. Morgan. Allstate Insurance (wholly owned by Sears, Roebuck): Chemical New York, Continental Illinois, First Chicago, and Crocker National. Mutual Life of New York:

Citicorp, Chemical New York, Bankers Trust, Crocker National, and Wells Fargo. New York Life: Citicorp, J. P. Morgan, Manufacturers Hanover (3), Chemical New York (2), and Crocker National. Continental Illinois: Chase Manhattan (2), Manufacturers Hanover (4), and J. P. Morgan.

The 1978 Staff Study recognized the magnitude of corporate concentration of control and heightened the importance and performance in the economy of very large bank holding companies that own and operate full- service banks. These very large bank holding companies, in their operations and performance through their enormous financial resources to make loans of substantial amounts of money and to buy and hold large blocks of securities, thereby form communities of interest and hold significant power to enable the concentration of control and to influence the shape and direction of American corporate and industrial growth while raising issues that involve unreasonable restraint of trade.

THE ENERGY INDUSTRY

There are eight large energy companies that provide the United States with a total energy package that include oil, gas, coal, nuclear, geothermal, and solar. The eight largest U.S. energy companies in 1976 consisted of (1) Exxon, (2) Texaco, (3) Mobil, (4) Standard (California), (5) Gulf, (6) Standard (Indiana), (7) Shell, and (8) Atlantic Richfield. From the standpoint of the concentration of control, the combined assets of the eight largest U.S. energy companies were equal to more than 10 percent of the total assets among the 130 major U.S. corporations in the Staff Study.[103]

Most significantly, there are three of the eight largest energy companies that had indirect interlocks with most of their major competitors, thus underscoring the unreasonable restraint of trade practices. Exxon is the largest energy company in the U.S. It indirectly interlocked with its leading competitors, thereby placing limits and restrictions on competition in such ways as being indirectly interlocked with Atlantic Richfield 4 times; with Mobil 6 times; with Shell 1 time; with Standard of California 6 times; with Standard of Indiana 2 times, and indirectly interlocked with Texaco 2 times. An assessment by the author of concentration in the structure of the energy industry has the effect of strengthening dominance when the recognition of the essential directorate affiliations of its major companies is ranked as one of the first orders of business. For example, as in the preceding analysis, the second largest energy company, Mobil, has indirectly interlocked with its sister competitors that included Exxon 6 times; indirectly interlocked with Atlantic Richfield 3 times; indirectly interlocked with Shell 3 times; indirectly interlocked with Standard of Indiana 3 times, and indirectly interlocked with Texaco 4 times.

The nation's third-largest energy company, Texaco, linked with Exxon 2 times; linked with Atlantic Richfield 2 times; and linked with Mobil 4 times.[104] Among the U.S. eight largest energy companies, concentration of control and interlocking directorates, in effect, the conduct or behavior conditions that foster reciprocal dealing or reciprocity, or the context of where their directors met, or the seats on the boards of directors, has the impact of making a contribution that could strengthen substantially the unreasonable restraint of trade, and alternatively, enable the increase in the degree of concentration of control in the economy.

For instance, Exxon had two directors, including its chairman, that had seats on the board of directors of Citicorp, a seat alongside the directors of Mobil, and a seat on the board of directors of Standard of California. Furthermore, other energy-related companies seated on the Citicorp Board of Directors were Halliburton (Brown & Root), Texas Eastern Transmission, General Electric, Stone and Webster, and Westinghouse Electric. The major energy-user companies that were seated on the Citicorp Board of Directors included General Motors, Ford, Monsanto, DuPont, Union Pacific, Southern Pacific, and U.S. Steel, while the major communications firms represented on the board of directors of Citicorp were AT&T, IBM, and RCA.[105]

The 1978 Staff Study, in its findings, reported that Exxon interlocked with Mobil and Texaco on the board of directors of AT&T. Also seated on that AT&T Board of Directors was a director from Continental Oil. There were major energy users that were seated on the board of directors of AT&T, such as General Motors, Chrysler, DuPont, Southern California Edison, Southern Railway, Union Carbide, and U.S. Steel.

The board of directors that provided the broadest opportunity for conducting business diplomacy or for holding a summit discussion on energy was Caterpillar Tractor, the heavy machinery manufacturer. Sitting on the board of directors of Caterpillar Tractor Company were directors from the major oil companies such as Exxon, Mobil, Shell, Standard of California, and Atlantic Richfield.

In addition, Chemical New York Bank, which had four major oil companies (directors from Exxon, Mobil, Atlantic Richfield, and Standard of Indiana) that met on Chemical New York's Board of Directors alongside the energy companies Amerada Hess and ARAMCO, was also directly interlocked with the Caterpillar Tractor Board of Directors. Moreover, AT&T, IBM, and U.S. Steel held seats on the board of directors of Caterpillar Tractor.[106]

MAJOR FINANCIAL INSTITUTIONS[107]

The 1978 Staff Study ascertained and emphasized the decision that the most significant assessment in the whole Staff Study has been limited by the representation of major financial institutions, most significantly, 16 major financial institutions based on their value to the country, size, and contribution for providing financial capital and investment services to the largest U.S. corporations, and that were selected out of the primary universe that included 130 major U.S. corporations that were evaluated. The 16 financial institutions included the following 10 banking companies and 6 insurance companies.[108]

Together these 16 major financial institutions represented a substantial share of revenue requirements, private capital that is available to major industrial and commercial enterprises, and in this connection, the 16 major financial institutions possess the power of dominance and the concentration of control that lead to the formation, shape, and direction of the growth, expansion, and diversification of corporations, international business, and multinational enterprises in the economy. These 16 major financial institutions, from the perspective of barriers to entry and freedom of entrants and competitiveness, also possess dominance roles in money management performance. In this regard, the 1978 Interlocking Directorates Staff Study addressed the market dominance role in money management by the inclusion and assessment of the periodical,*Institutional Investor* (August 1977)[109] and its announcement that the three hundred dominant money managers, primarily banking, insurance, and investment advisory companies, in effect, managed $822 billion in assets as of December 31, 1977, and reported the selected list of the ranking of the top three hundred money managers.

According to the August 1977 issue of *Institutional Investor*'s announcement, these preceding and selected twenty companies were ranked as leading in the ranking of the Top 300 Money Managers; most of the leading companies are included in the 1978 Staff Study.

The leading Pension Fund Managers that are reported and included by the *Institutional Investor* documentation are brought forth in this 1977 structure:[110]

Table 5.4.

Banking Companies	Insurance Companies
BankAmerica	Aetna Life & Casualty
Bankers Trust	Equitable Life & Casualty
Chase Manhattan	John Hancock
Chemical	Metropolitan Life
Citicorp	New York
Continental Illinois	Prudential
First Chicago	
Manufacturers Hanover	
J. P. Morgan	
Webster Bancorporation	

Table 5.5.

Name of Company[1]	*Total (Millions)*
Citibank	$26,103
Morgan Guaranty Trust	25,622
Prudential Insurance	25,218
Bankers Trust	24,000
Metropolitan Life	21,990
Mellon Bank	14,488
Scudder, Stevens & Clark	12,803
U. S. Trust	12,273
Equitable Life Assurance	11,990
Manufacturers Hanover	11,295
California Employees' Retirement	11,121
Aetna Life & Casualty	10,683
Harris Trust & Savings Bank	9,709
National Bank of Detroit	9,590
First National Bank of Chicago	8,894
Chase Manhattan Bank	8,552
Travelers Corp.	8,445
John Hancock Mutual	7,993
Continental Illinois	7,900
New York Life	7,695

[1] Ibid., p. 128

Table 5.6.

Name of Company	*Total (Millions)*
Bankers Trust	$19,000
Prudential Insurance	15,500
Morgan Guaranty Trust	15,013
Equitable Life Assurance	11,725
Citibank	11,680
Manufacturers Hanover Trust	7,293
National Bank of Detroit	7,192
First National Bank of Chicago	5,787
Scudder, Stevens & Clark	5,710
John Hancock Mutual Life	5,494
Harris Trust & Savings Bank	5,309
Lionel D. Edie	5,300
Travelers Corp.	4,523
First National Bank of Boston	4,444
Boston Co.	4,400
U. S. Trust	4,125
Continental Illinois National Bank	4,111
BankAmerica Corp.	3,596
Chase Manhattan	3,500
Chemical Bank	3,008

The interlocking directorates that are embedded in the 16 major financial institutions and included among the 130 major U.S. corporations included in the 1978 Staff Study, in fact, enable the potential for corporate concentration of control through the instrument of ownership and shareholding, and where the directors of the institutions that hold a substantial amount of stock or debt in a major company are seated on the boards of directors of that company or companies; they may be in a position to stop the company from acquiring financing, or they may be able to control that company's competition by the guarantee of necessary credit and required capital to the preferred company or companies. When large financial institutions interlock and enter linkages (partnerships) with major related industries, the result is the enabling of competitive advantage positions that permit the advancement of the concentration of control as well as the potential for increases in the barriers to entry and the strengthening of unfavorable restraint of trade.

INTERLOCKING DIRECTORATES IN THE AUTOMOTIVE INDUSTRY

General Motors, Ford Motor Company, and Chrysler have accounted for over 90 percent of the automobiles that are produced in the United States. These manufacturers, known as the "Big Three," make trucks, chassis for automotive equipment, and farm equipment along with making a large quantity of their own original and replacement parts. Moreover, the Big Three have finance subsidiaries that extend the retail and wholesale credit in the distribution of their products.

The main purpose of the 1978 Staff Study, involving the automotive industry, is to describe the pattern of indirect linkages that exist among competitors so that value judgments may be made as to the potential for concentration and anticompetitive behavior. In effect, the 1978 Staff Study is an examination of the direct interlocks between the Big Three automotive giants in order to determine or identify whether those close ties, or tying arrangements, if they exist, could lead to unreasonable restraint of trade and anticompetitive conduct and behavior.

The 1978 Staff Study provides documentation that the management of these major companies are represented by directors who are seated together on the boards of directors of America's largest companies, many of which are suppliers of materials and financial services, customers, and potential competitors in the automotive industry.

One other company deserves to be included in the automotive industry of the nation, and this company, namely International Harvester, is the only company among the 130 major U.S. corporations in the 1978 interlocking directorates staff study that has demonstrated the size, experience, and marketing potential for competing in the domestic automotive market. International Harvester also manufactures and sells trucks, chassis, farm equipment, and other automotive equipment and parts. The concentration of control facilitates the strengthening of substantial linkages in the automotive industry. In this connection, and in 1976, the directors of General Motors interlocked with directors of Chrysler on the boards of directors of AT&T, Chase Manhattan, National Detroit Corp., and the New York Stock Exchange.

There is an oligopolistic conduct and behavior pattern in operation through the General Motors linkages with Ford on the boards of directors of Citicorp, J. P. Morgan, and Procter & Gamble. In addition, General Motors connected with International Harvester on the boards of directors of Eastman Kodak, Santa Fe Industries, First Chicago, International Paper, and General Mills. Ford was found to link with Chrysler on the boards of directors of Hewlett-Packard, Utah International, and Burroughs Corp. The market structure in the automotive industry concerns vertical integration activity, and its accompanying strengthening of the concentration of control in addition to its expansion of indirect interlocking directorates. Vertical integration in the automotive industry is connected with backward integration into the ownership or control of distribution. For example, vertical integration and concentration of control involved the direct interlocks between General Motors and its customers, suppliers, and banking organizations, thus providing indications of automotive behavior and performance and unreasonable restraint of trade practices.

The General Motors Board of Directors in 1976 consisted of a director of both Canadian Pacific, which is a major customer, and Algoma Steel Co., which is a Canadian steel supplier. The board of directors of General Motors included the chairman of the board of J. P. Morgan, the dominant banker who also was a director of Santa Fe Industries, which is another large customer, and Bethlehem Steel, the second largest American steel supplier. Sitting on the board of directors of General Motors in 1976 also were the former chairman of the board of the Southern Co., who was a director of U.S Steel, the United States' largest steel supplier, and a director of Aluminum Company of America, which is one of the largest aluminum producers.[111]

Furthermore, the chairman of the board of AT&T, a major customer of General Motors, was on the board of directors of General Motors, and the AT&T chairman was also a director of Citicorp. On the General Motors Board of Directors were directors who represented the following major companies, among others: Chemical New York, Metropolitan Life, Mellon National, First Chicago Corp., Norfolk & Western, Kennecott Copper, and American Airlines. The president of General Motors, during this period, was a director of a leading insurance holding company, namely, Marsh & McLennan Companies., Inc., which was the second-largest stock voter in Chrysler Corp. as well as in American Broadcasting Co. and in CBS. General Motors is a vertically integrated company with inherent advantages of vertical integration, over nonintegrated companies. For instance, J. P. Morgan's Board of Directors provided a central forum for inevitable business diplomacy in the automotive industry. Seated on the board of directors of J. P. Morgan were directors of General Motors, Ford, DuPont, Bethlehem Steel, Burlington Northern, Western Electric (AT&T), General Electric, Union Carbide, Santa Fe Industries, Southern Railway, United Technologies, Bechtel, Coca-Cola, Corning Glass, Dow Corning, Ingersoll Rand, Phelps Dodge, and Owens Corning.[112]

Vertically integrated companies, as opposed to nonintegrated firms, are formations of important sources of the strengthening of monopoly power; they increase the barriers to freedom of entry on a fully integrated company basis. Moreover, vertical integration creates substantial capital market requirements that make it difficult, if not impossible, to raise the huge revenue requirements that are necessary for integrated company entry.

The direct interlocking between General Motors and its customers, suppliers, distribution facilities, and banking organizations could be considered unreasonable restraint of trade. As it pertains to the automotive industry, the Citicorp Board of Directors was concentrated with directors who represented companies involved in the automotive industry. Among these companies were AT&T, Bethlehem Steel, Boeing Co., Corning Glass, Cummins Engines, Dow Corning, DuPont, Exxon, Ford, General Electric, General Motors, Ingersoll Rand, Mobil, Pacar, Inc., Phelps Dodge, Pratt & Whitney (Canada), Southern Pacific, Standard Oil (California), United Technologies, and Westinghouse.

In the continued assessment of vertical integration competitive advantages and the potential for gaining monopoly power, it is important to recognize the Chase Manhattan Bank provided a common meeting ground for directors of General Motors, Chrysler, Armco Steel, Bendix, Cities Service, Exxon, Firestone Tire and Rubber, General Electric, Minnesota Mining and Manufacturing, Sperry Rand, U.S. Steel, and Union Carbide.

Multiple boards of directors provide for the conditions that facilitate the possibilities for business diplomacy, reciprocal dealing (reciprocity), and tying arrangements through which agreements may have been reached. For example, the National Detroit Corp., which is a bank holding company, has two directors holding seats on the board of directors of General Motors and another two directors were sitting on the board of directors of Chrysler. In addition, the National Bank of Detroit, which is a wholly owned subsidiary of National Detroit, is in fact the largest stock voter in General Motors. The National Bank of Detroit also has stockholdings in both Chrysler and Ford.

J. P. Morgan has a director sitting on the General Motors Board of Directors and a director sitting on the board of directors of Ford. AT&T has two directors on the General Motors Board of Directors, and it has one director on Ford's board of directors. Thus, all of these types and forms of linkages, connections, and strategic alliances set the stage for the maximization of opportunities of business and commercial ties, communication, and understanding between and among competitors.

The automotive industry suppliers and customers have placed emphasis on the importance of indirect board relationships and the concentration of control. In this regard, U.S. Steel was interlocked with Bethlehem Steel on the boards of directors of Citicorp and Metropolitan Life Insurance, as well as the General Motors Board of Directors. Furthermore, U.S. Steel connected with Chrysler on the board of directors of AT&T, Bell Canada, Chase Manhattan, General Foods, Manufacturers Hanover, and R. H. Macy & Co..

U.S. Steel met Ford on the boards of directors of Citicorp and Procter & Gamble. In fact, U.S. Steel indirectly interlocked with General Motors through American Airlines, AT&T, Chase Manhattan, Citicorp, Eastman Kodak, Hiram Walker, Kraft, Mellon National Bank, and Procter & Gamble.[113]

THE 1912 INTERLOCKING DIRECTORATES PROCESS OF CONCENTRATION STUDY

The Interlocking Directorates Process of Concentration Study, December 18, 1912, in essence, resulted from *Money Trust Investigation: Investigation of Financial and Monetary Conditions in the United States Under House Resolutions Nos. 429 and 504 Before a Subcommittee on Banking and Currency: Interlocking Directorates*, Washington, DC: Government Printing Office, 1913.

At its core, the 1912 Interlocking Directorates Process of Concentration Study represented essentially the processes of concentration of control in conjunction with the affiliations of eighteen dominant financial companies and institutions in the United States that form the "Agents of Concentration":[114]

- J. P. Morgan & Co., New York Astor Trust Co., New York
- First National Bank, New York New York Trust Co., New York
- Guaranty Trust Co., New York Blair & Co., New York
- Bankers Trust Co., New York Speyer & Co., New York
- National City Bank, New York Continental & Commercial—
- Kuhn, Loeb & Co., New York National Bank, Chicago

- National Bank of Commerce, New York First National Bank, Chicago
- Hanover National Bank, New York Illinois Trust & Savings Bank,
- Chase National Bank, New York Chicago
- Kidder, Peabody & Co., Boston and New York
- Lee, Higginson & Co., Boston and New York

The components practices that strengthen the processes of concentration of control have been heightened by and gained in importance as a result of the demonstrated interlocking directorates strategic and tacit global marketing management arrangements, interfirm linkages, affiliations, and connections.

For example, the 1912 Interlocking Directorates Investigation found that the affiliations of the selected companies as a group, namely, J. P. Morgan & Co., Guaranty Trust Co., Bankers Trust Co., First National Bank, and National City Bank, together, performed market dominance affiliations in the United States, and included were the following industry classifications: (1) banks and trust companies, (2) insurance companies, (3) transportation systems, (4) producing and trading corporations, and (5) public utility corporations. Specifically, in this connection, there were interlocking directorates such as:

a. Affiliations of J. P. Morgan & Co., First National Bank, National City Bank, the Guaranty Trust Co., Bankers Trust Co.[115]

J. P. Morgan & Co., First National Bank, National City Bank, Guaranty Trust Co., and Bankers Trust Co. together have —

- 118 directorships in 34 banks and trust companies
- 30 directorships in 10 insurance companies
- 105 directorships in 32 transportation systems
- 63 directorships in 24 producing and trading corporations
- 25 directorships in 12 public utility corporations

In all, 341 directorships in 112 corporations

b. Affiliations of J. P. Morgan & Co., Guaranty Trust Co., Bankers' Trust Co., First National Bank.

That J. P. Morgan & Co.., Guaranty Trust Co., Bankers Trust Co., and First National Bank, together, have 89 directorships in such banks and trust companies:

- 29 directorships in such insurance companies
- 78 directorships in such transportation systems
- 49 directorships in such producing and trading corporations
- 16 directorships in such public utility corporations.

In all, 261 directorships.

c. Affiliations of J. P. Morgan & Co., Guaranty Trust Co., Bankers Trust Co. That J. P. Morgan & Co., Guaranty Trust Co., and Bankers Trust Co., together, have:

- 78 directorships in such banks and trust companies
- 29 directorships in such insurance companies
- 64directorships in such transportation systems
- 44 directorships in such producing and trading corporations
- 14 directorships in such public utility corporations

In all, 229 directorships

NOTES

1. Ibid.
2. Ibid.
3. Ibid.
4. Ibid., p. 57.
5. Ibid., p. 67.
6. Ibid., pp. 86–87.
7. *Money Trust Investigation. Investigation of Financial and Monetary Conditions in the United States Under House Resolutions Nos. 429 and 504, before A Subcommittee of the Committee on Banking And Currency: Interlocking Directorates*, Washington, DC: Government Printing Office, December 16, 1918, Exhibit 134-C, pp. 1 and 27–28.
8. Bob Ivry, Bradley Keoun, and Phil Kuntz, "Secret Fed Loans Helped Banks Net $13 Billion," *Bloomberg Markets* magazine, November 27, 2011, p. 2 (http://www..bloomberg.com?news/orint/2011-11-28/secret-fed-loans-undisclosed-to-congr...11/28/2011).
9. The Financial Crisis Inquiry Report, *Final Report of the National Commission on the Causes of the Financial and Economic Crisis in the United States*, submitted by the Financial Crisis Inquiry Commission (FCIC) Pursuant to Public Law 111-21, January 2011, Official Government Edition, Chapter 20, p. 353.
10. Ibid.
11. Ibid.
12. Ibid., chapter 3, pp. 47–48.
13. Ibid., p. 50.
14. Ibid., pp.. 50–51.
15. Chairman Ben S. Bernanke, Board of Governors of the Federal Reserve System, Testimony: "Causes of the Recent Financial and Economic Crisis," before the Financial Crisis Inquiry Commission (FCIC), September 2, 2010, Washington, DC, P. 1 (www.federalreserve.gov/newsevents/testimony/bernanke20100902a.htm).
16. Ibid.

17. *Wall Street and the Financial Crisis: Anatomy of a Financial Collapse: Report and Appendix, before the Permanent Subcommittee on Investigations of the Committee on Homeland Security and Governmental Affairs*, U.S. Senate, 112th Congress, 1st Session, Volume 5, Part I, April 13, 2011, Washington, DC: U.S. Government Printing Office, 2011, p. 58.
18. Ibid., p.68.
19. Ibid., pp. 70–71.
20. Ibid., p. 192.
21. Ibid., pp. 193 and 211.
22. Ibid., pp. 194–195.
23. Ibid., pp. 195–196.
24. Ibid., p. 289.
25. Ibid., p. 307.
26. Ibid., p. 311.
27. Ibid., pp. 311–312.
28. Ibid., p. 312.
29. Ibid., p. 315.
30. Ibid.
31. Ibid., chapter 6, p. 377.
32. Ibid., p. 378.
33. Ibid., pp. 386–387.
34. Ibid., p. 387.
35. Ibid., p. 388.
36. Ibid., pp. 390–393.
37. Ibid., p. 416.
38. Ibid., pp. 416–417.
39. *Wall Street and the Financial Crisis: Anatomy of a Financial Collapse: Report and Appendix before the Permanent Subcommittee on Investigations of the Committee on Homeland Security and Governmental Affairs*, U.S. Senate, 112th Congress, 1st Session, Volume 5, Part II, April 13, 2011, Washington, DC: U.S. Government Printing Office, 2011, p. 866.
40. 95th Congress, 2d Session, U.S. Senate, Document No..95-99, *Voting Rights in Major Corporations: A Staff Study*, prepared by the Subcommittee on Reports, Accounting, and Management of the Committee on Governmental Affairs, January 1978, Washington, DC: U.S. Government Printing Office, 1978,, pp. 15–16.
41. *Wall Street and the Financial Crisis: Anatomy of a Financial Collapse, Report and Appendix Before the Permanent Subcommittee on Investigations of the Committee on Homeland Security and Governmental Affairs*, U.S. Senate, 112th Congress, 1st Session, Volume 5, Part I, April 13, 2011, p. 417.
42. The Financial Crisis Inquiry Report, *Final Report of the National Commission on the Causes of the Financial and Economic Crisis in the United States*, submitted by the FCIC, 2011, Chapter 8, pp. 130–131.
43. Ibid., p. 131
44. Ibid., pp. 132–133

45. *Wall Street and the Financial Crisis: Anatomy of a Financial Collapse: Report and Appendix, Before the Permanent Subcommittee on Investigations of the Committee on Homeland Security and Governmental Affairs, United States Senate*, 112th Congress, 1st Session, Volume 5, Part I, April 13, 2011, p. 418.
46. Ibid., pp. 419–421.
47. Ibid., pp. 433–434.
48. Ibid., pp. 438–440.
49. Ibid., p. 441.
50. Ibid., p. 440.
51. Ibid., p. 378.
52. Ibid., p. 378.
53. Ibid., (see also Ibid., pp. 387–388).
54. Ibid., pp. 448–449.
55. *The Financial Crisis Inquiry Report: Final Report of the National Commission on the Causes of the Financial and Economic Crisis in the United States*, Pursuant to Public Law 111-21, January 2011, p. 142.
56. Ibid., p. 142.
57. Ibid., p. 142.
58. Ibid.
59. Ibid., p. 143.
60. Ibid., p. 143.
61. Ibid., p. 143.
62. Ibid., p. 143.
63. *Wall Street and the Financial Crisis: Anatomy of a Financial Collapse: Report and Appendix, before the Permanent Subcommittee on Investigations of the Committee on Homeland Security and Governmental Affairs*, U. S. Senate, 112th Congress, 1st Session, Volume 5, Part I, April 13, 2011, p. 455.
64. Ibid., pp 463–464.
65. Ibid., pp. 464–465.
66. *The Financial Crisis Inquiry Report: Final Report of the National Commission on the Financial and Economic Crisis in the United States*, Pursuant to Public Law 111-21, January 2011, pp. 363–364.
67. Ibid., pp. 364–365.
68. Ibid., p. 365.
69. Ibid., pp. 298–299.
70. Ibid.
71. 96th Congress 2d Session, *Structure of Corporate Concentration: Institutional Shareholders and Interlocking Directorates Among Major U.S.* Corporations, Committee on Governmental Affairs, U.S. Senate, Volume 1, December 1980, Washington, DC: U.S. Government Printing Office, pp. 12–13.
72. Ibid., pp. 12–14.
73. Ibid., pp. 14–15.
74. Ibid., p. 15.
75. Ibid., p. 17.
76. Ibid., pp. 17–18.

77. Ibid., p. 18.
78. Ibid., p. 18.
79. Ibid., pp. 19–20.
80. Ibid., p. 20.
81. Ibid., pp. 20–21.
82. Ibid., p. 21.
83. Ibid., p. 21.
84. Ibid., p. 22.
85. Ibid., p. 22.
86. Ibid., p. 22.
87. Ibid., p. 22.
88. Ibid., p. 24.
89. Ibid., pp. 23–24.
90. Ibid., p. 25.
91. Ibid., p. 25.
92. Ibid., pp. 25–26.
93. 95th Congress, 2nd Session, U.S. Senate, Document No. 95-99, *Voting Rights in Major Corporations: A Staff Study*. prepared by the Subcommittee on Reports, Accounting and Management of the Committee on Governmental Affairs, January 1978, Washington, DC: U.S. Government Printing Office, 1978, p. 1.
94. Ibid., p. 285.
95. Ibid., pp. 3 and 260.
96. Ibid., p. 2.
97. Ibid., p. 2.
98. Ibid., p. 119.
99. Ibid., pp. 119–121.
100. Ibid., p. 122.
101. Ibid., pp. 123–124.
102. Ibid., p. 91.
103. Ibid., p. 91.
104. Ibid., p. 93.
105. Ibid., p. 93.
106. Ibid., pp. 120–121, Table 2E.
107. Ibid., pp. 119–121.
108. Ibid., p. 127.
109. Ibid., p. 127.
110. Ibid., pp. 75 and 78.
111. Ibid., p. 79.
112. Ibid., p. 79.
113. *Money Trust Investigation Investigation of Financial and Monetary Conditions in the United States Under House Resolutions Nos. 429 and 504 Before A Subcommittee of the Committee on Banking and Currency*, Washington, DC: U.S. Government Printing Office, 1913, p. 1.
114. Ibid., p. 27.
115. Ibid., p. 28.

6

Catastrophic and Calamity Consequences

In 1999, Congress passed the Gramm-Leach-Bliley Act (GLB Act) which repealed the prohibitions of Glass-Steagall Act, Affiliations Section 20, and Management Interlocks Section 32 of the Banking Act of 1933 (12 U.S. C. 377) and permitted the banks to affiliate with securities and insurance companies within financial companies. Under the provisions of the Gramm-Leach-Bliley Act (1999), a bank holding company that has met the qualifications to become a financial holding company may participate in a broader range of financially connected activities, including full-scope securities underwriting and dealing, insurance underwriting, and sales, and merchant banking. A bank holding company has the meaning that is given the term in Section 2 of the Bank Holding Company Act of 1956, such as any company that directly or indirectly owns, controls, or has the power to vote 25 percent or more of any class of the voting shares of a bank; controls in a manner the election of a majority of the directors or trustees of a bank; or is found to exercise a controlling influence over the management or policies of a bank. A bank holding company must obtain the approval of the Federal Reserve before acquiring more than 5 percent of the shares of an additional bank or bank holding company. All bank holding companies must file certain reports with the Federal Reserve System.

The concentration in (1) the Structure of Corporate Concentration Institutional Shareholders Among 122 Major U.S. Corporations, its convergence with (2) the concentration in Voting Rights in 130 Major U.S. corporations, in consolidation with (3) the Investigation of Concentration of Economic Power Distribution of Ownership in the 200 largest corporations—Temporary National Economic Committee (TNEC), and in addition to the contribution of concentration produced by The Report of the Committee Appointed

Pursuant to House Resolutions 429 and 504 to Investigate the Concentration of Control of Money and Credit, thereby and in total, have formed the creation, consolidation and centralization of the market structure of mega aggregate concentration of ownership and control.

In recognition of the significance of the Gramm-Leach-Bliley Act, passed by Congress in 1999, mega aggregate concentration of control through interlocking directorates, financial ties, structure of corporate concentration-institutional shareholders, together with interlocking directorates among the major U.S. corporations supported and bound by interlocking directorates in the Money Trust Investigation of Financial and Monetary Conditions in the United States Under House Resolutions Nos. 429 and 504, as a result the mega aggregate concentration of ownership and control has created and driven the establishments of the substantial increase in the acceleration rate and in the gigantic volume of transactions utilization in the global over-the-counter (OTC) derivatives accumulation in excess of the demand for U.S. dollars; aimed and substituted for, and indeed, targeted the replacement of the demand for U.S. Money Stock Measures M1 and M2. Thus, the tragic event, strategy of ruin, catastrophic consequences and its determinants based on the mega aggregate concentration of ownership and control through interlocking directorates financial and management ties among the largest corporations and major U.S. corporations.

The definition and significance of the U.S. Money Stock Measures[1] (in billions of dollars) is presented from the Federal Reserve Statistical Release, H.6 (508), January 2, 2009. 1. M1 consists of (1) currency outside the U.S. Treasury, Federal Reserve Banks, and the vaults of depository institutions; (2) traveler's nonbank issues; (3) demand deposits at commercial banks (excluding those amounts held by depository institutions, the U.S. government, foreign banks and official institutions) less cash items in the process of collection and Federal Reserve float; and (4) other checkable (OCDs), consisting of negotiable order of withdrawal (NOW) and automatic transfer service (ATS) accounts at depository institutions, credit share draft accounts, and demand deposits at thrift institutions. Seasonally adjusted M1 is constructed by summing currency, traveler's demand deposits, and OCDs, each seasonally adjusted separately. 2. M2 consists of M1 plus (1) savings deposits (including money market deposit accounts): 2) small-denomination time deposits (time deposit amounts of less than $100,000), less individual retirement account (IRA) and Keogh balances at depository institutions; and (3) balances retail money market mutual funds, less IRA and Keogh balances at money market mutual funds. Seasonally adjusted M2 is constructed by summing savings deposits, small-denomination time deposits, and retail money funds, each seasonally adjusted separately, and adding this result to seasonally adjusted M1.

Table 6.1. Table B-69 Money Stock and Debt Measures, 1972-2011 (Averages of Daily Figures, Except Debt end-of-period Basis: Billions of Dollars, Seasonally Adjusted)

	M1	*M2*	*Debt*	*Percent Change*		
Year and Month	*Sum of Currency, Demand Deposits, Travelers Checks and Other Checkable Deposits (OCDs)*	*M1 Plus Retail MMMF Balances Savings Deposits (Include MMDAs and Small Time Deposits)*	*Debt of Domestic Nonfinan. Sectors*	*From Year or 6 Mos. Earlier From Prev. Period Debt*		
2000	1,087.8	4,911.6	18,165.4	–3.1	6.1	5.0
2001	1,182.8	5,424.9	19,297.5	8.7	10.5	6.3
2002	1,220.6	5,768.7	20,716.1	3.2	6.3	7.4
2003	1,307.0	6,058.6	22,443.8	7.1	5.0	8.1
2004	1,376.7	6,404.4	25.264.5	5.3	5.7	9.2
2005	1375.0	6,667.9	27,588.5	–.1	4.1	9.2
2006	1,367.2	7,059.7	29,979.4	–.6	5.9	8.7
2007	1,375.4	7,484.5	32,525.7	.6	6.0	8.5
2008	1,606.9	8,231.7	34,463.7	16.8	10.0	6.0
2009	1,697.8	8,514.2	35,430.4	5.7	3.4	3.0
2010	1,840.3	8,796.4	36,931.6	8.4	3.3	4.1
2011	2,173.9	9,640.1		18.1	9.6	

An essential, detailed and structured documentation of Money Stock Measures M1 and M2 are included in Table 6.1: Money Stock and Debt Measures, 1972–2011 from the perspective of MONEY STOCK, CREDIT, and FINANCE.[2]

Note: For further information on the composition of M1 and M2, see the H6 release of the Federal Reserve Board. The Federal Reserve no longer publishes the M3 monetary aggregate and most of its components. Institutional money market mutual funds is published as a memorandum item in H.6, and the component one large-denomination time deposits is published in other FRB releases.

The global over-the-counter (OTC) derivatives accumulation in excess of the demand for dollars and aggrandizement targeted the U.S. money stock measures M1 and M2, structured in the preceding Table 6.1, for decrease in demand for dollars, diminution, and for substitution and replacement through the gigantic volume of trading in global OTC derivatives accumulation contracts that are in excess of the demand for dollars. The global OTC derivatives position market and the gigantic global OTC derivatives accumulation and galvanization, jointly, soared to the huge volume of trading in derivatives transactions that reached the level of $683.7 trillion notional/contract amount in the financial and economic crisis year 2008 an continued the phenomenal rise of the volume of trading in global OTC derivatives to $706.8 trillion notional/contract amount in 2011, which is far in excess above the trend of the 1995 low of $55.7 trillion (March 3, 1995); $94.0 trillion in 2000; and the $507.9 trillion in the 2007 other financial and economic crisis year; discussed in chapter 4.

This huge global OTC derivatives accumulation in excess of the demand for dollars urgency and significance centers on the U.S. Federal Reserve Seal of Bank Approval of Large Banks, Major Banks, Bank Holding Companies, and Financial Bank Holding Companies, such as (a) the top 50 bank holding companies (BHCs) as of June 30, 2012, National Information Center A Repository of Financial Data and Institution Characteristics Collected by the Federal Reserve System; the composition of the top-50 panel is based on consolidated assets (FFIEC.gov/nicpubweb/nicweb/top50form.aspx); (b) Federal Reserve Bank: Large Commercial Banks, Federal Reserve Statistical Release. Insured U.S.—Chartered Commercial Banks that have consolidated assets of $300 Million or more, Ranked by Consolidated Assets as of March 31, 2012 (http://www.federalreserve.gov/releases/1br/current/default.htm); (c) List of Financial Holding Companies as of May 18, 2012, Board of Governors of the Federal Reserve System. Under the Bank Holding Company Act, the bank holding companies may elect to be financial holding companies whose elections to become or be treated as financial holding companies are effective, .(http://www.federalreserve.gov/bankin foreg/fhc.htm); (d) the Primary Dealers List, Federal Reserve Bank of New York, Primary Dealers serve as trading counterparties of the New York Federal Reserve Bank in its implementation of monetary policy. This role includes the obligations to (i) participate consistently to carry out U.S. monetary policy pursuant to the direction of the Federal Open Market Committee (FOMC), (ii) provide the New York Fed's Trading Desk with market information and analysis that are helpful in the formulation and implementation of monetary policy. The Primary Dealers are also required to participate in all auctions of U.S. government debt and to make reasonable markets for the New York Fed when it transacts on behalf of its Foreign Official Account Holders, (newyorkfed.org/markets/pridealers_current.html): (e) Reverse Repo Counterparties List, Federal Reserve Bank of New York, prepared for the potential need to conduct large-scale reverse repurchase agreement transactions; the Federal Reserve Bank of New York is developing arrangements with an expanded set of counterparties with whom it can conduct thee transactions. These counterparties are in addition to the existing set of Primary Dealers counterparties with whom the Federal Reserve can already conduct reverse repurchase agreements, (www.newyorkfed.org/markets/expanded_ counterparties.html); (f) Nineteen (19) Bank Holding Companies, Board of Governors of the Federal Reserve System Comprehensive Capital Analysis and Review: Objectives and overview (CCAR), March 18, 2011.

These nineteen bank holding companies also participated in the 2009 SCAP. Capital is central to a bank holding company's ability to absorb unexpected losses and continue to lend to creditworthy businesses and consumers. The Federal Reserve expects large, complex bank holding companies to hold sufficient capital to maintain access to funding, to continue to serve as credit intermediaries, to meet their obligations to creditors and counterparties, and to continue operations, even in an adverse environment.

Table 6.2. The 19 Bank Holding Companies Participating In the Board of Governors of the Federal Reserve System Comprehensive Capital Analysis and Review (CCAR): Objectives and Overview, March 18, 2011

Ally Financial, Inc.	Bank of America Corporation
American Express Company	The Bank of New York Mellon
BB&T Corporation	Citigroup, Inc.
Capital One Financial	The Goldman Sachs Group, Inc.
Fifth Third Bancorp.	Keycorp
JP Morgan Chase & Co.	Morgan Stanley
Metlife, Inc.	Regions Financial Corporation
The PNC Financial Services Group, Inc.	U. S. Bancorp
State Street Corporation	Wells Fargo & Company
Sun Trust Banks, Inc.	

The significance of the Federal Reserve System series of approval as the most salient aspects of the structure among which are included the following segments, (a) The top 50 bank holding companies, (b) Federal Reserve Bank: Large Commercial Banks, (c) Federal Reserve Bank List of Financial Holding Companies, (d) FRB New York Primary Dealers, (e) FRB New York Reverse Repo Counterparties List, and (f) the Federal Reserve Bank 19 Bank Holding Companies Comprehensive Capital Analysis and Review, in essence and in toto, failed to include and acknowledge what the 1978 VOTING RIGHTS IN MAJOR U.S. CORPORATIONS A Staff Study uncovered and reported the following finding, "who controls the major voting rights in the dominant banks," and provided the answer, "the principal stockvoter in large banks are large banks."

This 1978 VOTING RIGHTS IN MAJOR U.S. CORPORATIONS A Staff Study continued and gave prominence to the fact that under Morgan Guaranty's dominance position, the Morgan Guaranty Trust Company of New York is stockvoter number one (1) in four of its New York sister banks—Citicorp, Manufacturers Hanover Corp., Chemical New York Corp. and Bankers Trust New York Corp.—as well as BankAmerica Corp. Moreover, Citicorp is stockvoter number one (1) in Morgan Guaranty's parent holding company, J. P. Morgan & Co., thus, who controls the major voting rights in the dominant banks, the principal stockvoter in large banks are large banks.The 1978 VOTING RIGHTS IN MAJOR U.S. CORPORATIONS A Staff Study disclosed Morgan Guaranty Trust Co. of New York towers above all other investors and is among the top five identified stockvoters in almost half (56) of the 122 corporations studied. The Morgan Guaranty Trust Company of New York is stockvoter number one (1) in almost one-fourth of the corporations.

The mega aggregate concentration of control, voting rights in major U.S. corporations and interlocking directorates, in total, conducted international business diplomacy with the Bank for International Settlements (BIS) sixty members central bank, as well as the Federal Reserve System and the Federal Reserve Bank of New York, perpetuated the galvanization and maximization of global OTC derivatives accumulation in excess of the demand for dollars and huge volume of trading transactions performed under the condition of heightened and strengthened barriers to entry, limitation of competition and unreasonable restraint of trade practices.

FEDERAL RESERVE SYSTEM INJECTIONS AND RECIPROCAL CURRENCY ARRANGEMENTS (SWAP LINES) ACTIVITY

The measurement of the catastrophic and calamity consequences of the financial and economic crisis can be assessed also from the perspective of the

Federal Reserve System Injections, Interventions and Reciprocal Currency Arrangement (Swap Lines) Activity. For example, the Federal Reserve System Federal Open Market Committee (FMOC), with appropriate policy accommodation, has continued purchasing additional agency mortgage-backed securities at a pace of $40 billion per month and longer-term Treasury securities at a pace of $45 billion per month for a total of $85 billion per month. (http://www.federalreserve.gov/newsevents/press/monetary/210301.30a.htm), January 30, 2013.

The Swap arrangements were introduced to address stresses in the U.S. dollar funding in overseas markets. These difficulties were adding materially to pressures in funding and credit markets in the United States and abroad. The dollar liquidity lines were designed to improve liquidity conditions in the United States and foreign financial markets by providing foreign central banks with the capacity to deliver U.S. dollar funding to institutions in their jurisdictions during times of market stress. The Federal Reserve provided U.S. dollars to a foreign central bank. At the same time, the foreign central bank provided the equivalent amount of funds in its currency to the Federal Reserve, based on the market exchange rate at the time of the transaction. The parties agreed to swap back these quantities of their two currencies at a specified date in the future, which was the next day or as far ahead as three months, using the same exchange rate as in the first transaction. Because of the terms of this second transaction were set in advance, the fluctuations in exchange rates during the interim did not alter the eventual payments.

On September 18, 2008, the Federal Reserve, the Bank of Canada, the Bank of England, the European Central Bank (ECB), the Bank of Japan, and the Swiss National Bank announced coordinated measures designed to address the continued elevated pressures in the U.S. dollar short-term funding markets. These measures, together with other actions taken during the last few days by individual central banks, were designed to improve the liquidity conditions in global financial markets. The central banks continued to work together closely, and will take appropriate steps to address the ongoing pressures, (http://www.federalreserve.gov/newsevents/press/monetary/20080918a.htm). The Federal Reserve actions through the Federal Open Market Committee (FOMC), on September 18, 2008, authorized a $180 billion expansion of its temporary reciprocal currency arrangements (swap lines). This increased capacity was available to provide dollar funding for both term and overnight liquidity operations by the other central banks. The Federal Open Market Committee (FOMC) authorized increases in the existing swap lines with the European Central Bank (ECB) and the Swiss National Bank. These larger facilities support the provision of U.S. dollar liquidity in amounts of up to $110 billion by the ECB, an increase of $55 billion, and up to $27 billion by the Swiss National Bank, an increase of $15 billion.

In addition, new swap facilities were authorized with the Bank of Japan, the Bank of England and the Bank of Canada. These facilities upported the provision of United States dollar liquidity in amounts f up to $60 billion by the Bank of Japan, $440 billion by the Bank of England, and $10 billion by the Bank of Canada. In 2008, and with special reference to "Lending to Primary Dealers," the Federal Reserve created special facilities that provided loans of cash and securities to primary dealers who are the securities-broker dealers that have a trading relationship with the Federal Reserve Bank of New York. These programs were closed on February 1, 2010. Apart from these special facilities, the Federal Reserve operates an ongoing overnight securities lending program to provide a temporary source of Treasury and agency securities to promote the smooth clearing of the Treasury and agency securities market, (http://www.federalreserve.gov/monetarypolicy/bst_lendingprimary.htm). On April 6, 2009, the Federal Reserve, the Bank of England, the European Central Bank, the Bank of Japan, and the Swiss National Bank announced swap arrangements that would enable the provision of foreign currency liquidity by the Federal Reserve to U.S. financial institutions. The euro, yen, sterling, and Swiss francs were provided to the Federal Reserve via these additional swap agreements with the relevant central banks. Furthermore, the central banks continued working together and taking steps as appropriate to foster stability in global financial markets, (http://www.federalreserve.gov/newsevents/press/monetary/20090406a.htm).

The Federal Reserve Chairman Ben S. Bernanke delivered his testimony, "Causes of the Recent Financial and Economic Crisis," before the Financial Crisis Inquiry Commission, Washington, DC, September 2, 2010, affirming the financial and economic crisis catastrophic consequences through his testimony statement that too-big-to-fail financial institutions were both a source (though by no means the only source) of the crisis and among the primary impediments to policy makers' efforts to contain the crisis. Chairman Bernanke stated that with more than $1 trillion in subprime mortgages outstanding, the potential for losses on these loans was large in absolute terms; in contrast and judged in relation to the size of global financial markets, the prospective subprime mortgages losses were clearly not large enough on their own to account for the magnitude of the financial and economic crisis.

Most significantly, Chairman Ben Bernanke confided that the system's vulnerabilities (the structured weaknesses in the financial system and in regulation and supervision that propagated and amplified the initial shocks), together with gaps in the government's crisis-response toolkit, were the principal explanations of why the crisis was so severe and had such devastating effects on the broader economy, (http://www.federalreserve.gov/newsevents/tectimony/bwernanke20100902a.htm).

Although the subprime mortgage losses were the most prominent trigger (the particular events or factors that touched of the crisis) of the crisis, the Federal Reserve Chairman Bernanke continued, they were by no means the only one; another but less well known triggering event was a sudden stop in June 2007 in syndicated lending to large, relatively risky corporation borrowers. In his testimony before the Financial Crisis Inquiry Commission (FCIC), Chairman Bernanke addressed conditions prior to the crisis and announced that, leading up to the crisis, the shadow banking system (shadow banks are financial entities other than regulated depository institutions such as commercial banks, thrifts and credit unions that serve as intermediaries to channel savings into investment), as well as some of the largest global banks, had become dependent on various forms of short-term wholesale funding. An increase in the risk of a run induces financial firms to hoard liquidity, for example, by shifting asset holdings into highly liquid securities such as Treasury securities.

CALAMITY CONSEQUENCES

The financial crisis, economic collapse, and monetary catastrophic and calamity consequences from 2007–2008 to 2012–2013 emanated from a bounded rationality fallacy, first and foremost, that the pervasive gigantic global OTC derivatives accumulation in excess of the demand for dollars aggrandizement must become the central components of the monetary aggregates and take the place of, substitute for and reduce the demand for U.S. Money Stock Measures M1 and M2, which is interwoven and connected with the dominance driven by mega aggregate concentration of ownership and control and interlocking directorates.

The global OTC derivatives accumulation are not U.S. Money Stock Measures M1 and M2. In fact, the huge global OTC derivatives contracts accumulation and financial products possess and consist of underlying variables that overvalued items, have inherent and numerous deficiencies; have broad-scale declines in underwriting standards; increased reliance on vague, misleading, concealed, cloaked and opaque credit instruments; have illiquid transactions that require the buying of U.S. Treasury securities in order to possess liquidity; global OTC derivatives accumulation lack reliable credit ratings and lack rewarding compensation for risk taking.

The major corporations, banks, bank holding companies and financial holding companies, which have engaged in the maximization of the volume of trading in global OTC derivatives accumulation, in fact, have experienced substantial capital depletion due to the losses and writedowns and their balance sheet clogged by the deterioration, harmful and the declining utilization and declining demand for OTC derivatives accumulation contracts. The inherent inequities and the loss of liquidity in OTC derivatives accumulation have been directly connected with substantial increases in the rising average costs of companies along with the heightened level of OTC derivatives based liabilities over assets in addition to widespread diseconomies of scale and persistent diminishing returns, thereby to the unraveling of mega aggregate concentration of ownership and control, weakening of interlocking directorates linkages, and cresting tragic events—calamity consequences of the financial and economic crisis from 2007–2008 to 2012–2013. The necessity for the enforcement of the Sherman Act has gained additional momentum and urgency because of the huge global over-the-counter (OTC) derivatives accumulation market expansion and, simultaneously, the growth in the concentration of control by exchanges that trade in global OTC derivatives, such as specifically, the Intercontinental Exchange, Inc. (ICE), proprietary trading, hedge funds and private equity funds. The Intercontinental Exchange, Inc. (ICE), and sustained assault on the Money Stock Measures M1 and M2, operates four regulated futures Exchanges, (a) ICE Futures Europe, (b) ICE Futures Canada, (c) the Chicago Climate Exchange, and (d) ICE Futures U.S. In addition, ICE also owns and operates five derivatives clearinghouses, (1) ICE Clear S, a derivatives clearing organization under the U.S. Commodity Exchange Act, and located in New York and serving the markets of ICE Futures U.S., (2) ICE Clear Europe, a recognized clearinghouse and derivatives clearing organization located in London that serves ICE Futures Europe, ICE's OTC energy markets and operates as ICE's European credit default swaps (CDS) clearinghouse, (3) ICE Clear Canada, a recognized clearinghouse located in Winnipeg, Manitoba, that serves the markers of ICE Futures Canada, (4) The Clearing Corporation, a U.S. derivatives clearing organization, and (5) ICE Clear Credit, a U.S. based CDS clearinghouse.[3]

Having already become the management of a diverse set of exchanges (perhaps including the New York Stock Exchange) and clearinghouses based in the U.S., Europe, and Canada, the Intercontinental Exchange, Inc. (ICE) has gained market dominance in the trading of global OTC derivatives accumulation in excess of the demand for dollars, an unreasonable restraint of trade. Chairman Gary Gensler, U.S. Commodity Futures Trading Commission (CFTC) stated on December 12, 2012, in his testimony before the U.S. House Financial Services Subcommittee on Capital Markets and Government Sponsored Enterprises, that the financial crisis cost eight million American jobs, millions of people lost their homes, and thousands of businesses closed their doors—in part because of the unregulated swaps market.

In the following significant literature, namely, the *Wall Street and the Financial Crisis: Anatomy of a Financial Collapse*; *The Financial Crisis Inquiry Report—Final Report of the National Commission on the Causes of the Financial and Economic Crisis in the United States*, together disclose in effect the seeds of catastrophic and calamity consequences in the U.S. economy and global ramifications.

Table 6.3: (B. 100) Balance Sheets of Households and Nonprofit Organizations, which is a compilation of Flow of Funds Accounts of The United States, Board of Governors of the Federal Reserve System, particularly is a representation of economic deep distress marked by great loss and suffering through the huge decreases in FL155035015 Households, FL153165105 Home Mortgages and FL155035065 Owner's Equity in Household Real Estate, 2005 to 2011; including other sources from the Flow of Funds Account (FFA) such as FFA Coded Tables, December 6, 2012; FFA Coded Tables, December 8, 2011; and FFA Coded Tables, June 8, 2006.

Notes: (i) Sector includes domestic hedge funds, private equity funds, and personal trust; (ii) At market value; all types of owner-occupied housing including farm houses and mobile homes, as well as second homes that are not rented, vacant homes for sale, and vacant land; (iii) includes loans made under home equity lines of credit secured by junior liens, shown on Table L.218, Line 23; and (iv) Line 4: FL155035015 Households, Less Line 33: FL153166000 Home Mortgages.

In Table 6.3: Balance Sheets of Households and Nonprofit Organizations, the owners' equity in household real estate in 2005 is $16,188.0 billion that dropped substantially in 2011 to a new low level of $6,452.5 billion reflecting the huge loss in the amount of $4,735.5 billion representing a major level of economic distress and misfortune experience for people who are affected.

Table 6.3. (B.100) Balance Sheet of Households and Nonprofit Organization, (i) Billions of Dollars; Amounts Outstanding End of Period, Not Seasonally Adjusted[1]

Year	*FL155035015 Households (ii)*	*FL153165105 Home Mortgages (iii)*	*FL155035065 Equity in Household Real Estate (iv)*
2002	13,671.2	5,844.0	7,827.2
2003	15,057.7	6,679.5	8,378.1
2004	17,209.9	7,604.5	9,605.4
2005	19,870.9	8,682.9	11,188.0
2007	20,870.3	10,540.4	10,329.8
2008	17,560.0	10,609.1	7,050.9
2009	17,151.3	10,360.9	6,790.3
2010	16,603.1	9,941.9	6,661.2
2011	16,165.2	9,712.7	6,452.5

1

When being examined through the structure of the following Table 6.4: Employment, Unemployment, Status of the Labor Force (Thousands of Persons 16 Years of Age and Over), the calamity consequences of the huge global over-the-counter (OTC) derivatives accumulation are found in the results of the trends specified in the data.

During the period that covered 2007 to 2011, the civilian employment declined from the level of 146,047,000 employees in 2007 to the lower level, at 139,869,000 employees in 2011, a decrease of 6,178,000 persons. Over the same period the number of unemployed persons was 7,078,000 in 2007, increasing to a level of 13,747,000 in 2011, reflecting an overall increase in the number of persons who were unemployed in the overall number of 6,669,000 unemployed persons.

In the category of "Not in the Labor Force" which is distinctly different from the "labor force" that is made up of the employed and unemployed, the remainder or those persons who have no jobs and are not looking for a job when the survey was taken, are counted as "not in the labor force." Many of the persons who are classified as "not in the labor force" are going to school or are retired. Family responsibilities keep others out of the labor force; individuals must indicate that they currently want a job, have looked for work in the last twelve months, and are available for work.

In Table 6.4: Employment, Unemployment, Status of the Labor Force (Thousands of Persons 16 Years of Age and Over) there were 78,743,000 persons in 2007 included in the category of "Not in the Labor Force" and, in 2011, there were 86,001,000 persons were found to be "Not in the Labor Force," indicating the increase of 7,258,000 persons.

The unemployment rate nearly doubled over the same period as demonstrated by the rising unemployment rate from 4.6 percent in 2007, to 5.8 percent in 2008, with a jump in the unemployment rate to 9.3 percent in 2009,

Table 6.4. Employment, Unemployment, Status of the Labor Force[1] (Thousands of Persons 16 Years of Age and Over)

Period	*Civilian Employment Total*	*Unemployment Total*	*Not in Labor Force*	*Unemployment Rate Percent*
2006	144,427	7,001	77,387	4.6
2007	146,047	7,078	78,743	4.6
2008	145,362	8,924	79,501	5.8
2009	139,877	14,265	81,659	9.3
2010	139,064	14,825	83,041	9.6
2011	139,869	13,747	86,001	8.9

Table 6.5. (Table 711) People Below Poverty Level and Below 125 Percent of Poverty Level[1] 1980 to 2009 (29,272 represents 29,272,000. People as of March of the following. Based on Current Population Survey. Annual Social and Economic Supplement (ASBC); see text, this section. Section 1, and Appendix III. For data collection changes over time, see http://www.census.gov/www/income/data/historical/history.html)

Year	*Number of Persons Below Poverty (1,000) All Races*	*Percent of Persons Below Poverty All Races*	*Below 125 Percent of Poverty Level*	
			Number (1,000)	*% of Total Population*
2005	36,950	12.6	49,327	10.8
2006	36,460	12.3	49,688	10.0
2007	37,276	12.5	50,876	17.0
2008	39,829	13.2	53,805	17.8
2010	46,216	15.3		
2011	48,452	15.9		

then increasing to the high unemployment rate of 9.6 percent in 2010 before leveling off at the 8.9 percent rate of unemployment in 2011 that numbered 13,747,000 unemployed persons in 2011.

The calamity consequences spread their harmful and economically distressed impact and included people who live below the poverty level and below 125 percent of the poverty level, which is specified in Table 6.5—People Below Poverty Level and Below 125 Percent of Poverty Level 1980 to 2009.

Between 2005 and 2011 in Table 6.5, the number of people below poverty, the percent of persons below poverty, and below 125 percent of poverty level, together, indicate the persistent increase in the number of people in poverty in the United States. For example, in 2005, the number of persons below poverty was 36,950,000. There were 36, 460,000 persons below poverty in 2006; 37,276,000 in 2007; 39,829,000 in 2008; 43,569,000 in 2009; and in 2010 and 2011, respectively, the number of persons below poverty had increased substantially to 46,15,956 and the high level of 48,452,035.

A different perspective and assessment of the impact of the financial and economic crisis and collapse tragic effects was documented by the Federal Reserve in its article entitled, "Changes in U.S. Family Finances from 2007 to 2010: Evidence from the Survey of Consumer Finance," Federal Reserve Bulletin, June 2012, Vol. 98, No. 2,[4] which is referred to or called the Federal Reserve Board's Survey of Consumer Finances (SCF) for 2010.

Some insights into the changes in family income and net worth since the previous 2007 survey, according to the Federal Reserve Board's Survey of Consumer Finances (SCF) for 2010, detailed that, covering the 2007–2010 period, the median value of real (adjusted for inflation) family income before taxes fell 7.7 percent; also the median income had fallen slightly during the previous three-year period.

More specifically, the decline in median income was broad-scale across demographic groups of persons. The decline in median income was most pronounced among more highly educated families, families headed by persons aged less than fifty-five, and families who live in the South and West regions.

The real mean income fell by 11.1 percent across all families, which is even more than median income in the recent period. This decline in mean income was even more widespread than the decline in the median income, wherein virtually all demographic groups of persons experienced a decline between 2007 and 2010.[5]

In addition, the Federal Reserve Board's Survey of Consumer Finances (SCF) for 2010 found that, although the declines in the values of financial assets or business were important factors for some families, the decreases in the median net worth appear to have been driven most strongly by a broad collapse in house prices. Moreover, if primary residences and the associated mortgage debt are excluded, then the median of families' net worth is reduced substantially from $126,400 to $42,300 in 2007, and from $77,300 to $29,800 in 2010.[6] The broad collapse in house prices is reflected in the patterns of change in net worth across demographic groups to varying degrees, depending on the rate of home ownership and the proportion of assets that are invested in housing. The decline in median net worth was especially large for families in groups where housing was a larger share of assets and where, for example, the families that are headed by someone 35 to 44 years old (median net worth fell 54.4 percent) and families in the West region (median net worth fell 55.3 percent).[7]

In this connection, the origin of calamity consequences deepened in scale and scope as early as the period between 1990 and 2004 when homeownership rates in the United States increased rapidly from 64 percent to 69 percent, representing the highest level in 50 years. In this regard, the Federal National Mortgage Association (Fannie Mae) and the Federal Home Loan Mortgage Corporation (Freddie Mack), two government-sponsored enterprises (GSEs) promoted homeownership essentially by providing a secondary market for home mortgages.

These two government-sponsored enterprises created that secondary market through the purchase of loans from lenders, securitizing them, ensuring a guarantee that they would make up the cost of any securitized mortgage that defaulted, and selling the resulting mortgage backed securities (MBS) to investors, thereby leading many to believe that the securities had the implicit backing of the federal government and to view the securities as very safe investments, and leading the investors around the world to buy them.

In its findings, the *Wall Street and the Financial Crisis: Anatomy of a Financial Collapse*, Report and Appendix, Before The Permanent Subcommittee on Investigations of the Committee on Homeland Security and Governmental Affairs, U.S. Senate, April 13, 2011 reported that the existence of this government-sponsored enterprise (GSE) secondary market had encouraged the lenders to originate more loans, since they could easily sell them to the government-sponsored enterprises and use the profits to increase their lending.[8]

Fannie Mae and Freddie Mack purchased larger quantities of higher risk loans and encouraged their proliferation. Fannie Mae alone purchased billions of dollars in high home loans and loans with subprime characteristics during 2005 and 2007. By the middle of 2009, Fannie Mae reported an unpaid principal balance of $878 billion for its loans with subprime characteristics, which represented about one third of its total portfolio of $2.7 trillion.

The *Wall Street and the Financial Crisis: Anatomy of a Financial Collapse*, Report and Appendix, Before the Permanent Subcommittee on Investigations of the Committee n Homeland Security and Governmental Affairs, U.S. Senate, April 13, 2011, in its conclusion, stated that the most immediate trigger to the financial crisis was the July 2007 decision by Moody's and Standard & Poor's (S&P) to downgrade hundreds of residential mortgage-backed securities (RMBS) and collateralized debt obligations (CDO) securities. The firms took this action because, in the language of one S&P senior analyst, the investment grade ratings could not hold or be maintained.

The credit rating agencies admitted that the "emperor had no clothes" and acknowledged that RMBS and CDO securities contained high risk, poor quality mortgages that were not safe investments and were going to incur losses. The investors stopped buying, then the value of the RMBS and CDO securities fell, and the financial institutions around the world were suddenly left with unmarketable securities whose value had plummeted. The financial crisis was on.[9]

Credit rating agencies and credit ratings are impermissible under the Dodd-Frank Wall Street Reform and Consumer Protection Act of 2010. All U.S. Federal agencies must remove references to, and requirements of, reliance on credit ratings from their regulations and replace them with appropriate alternatives for evaluating creditworthiness under the Dodd-Frank Act.

In order to overcome the tragic events produced by the credit rating agencies activity during the financial crisis, there have been created "Risk-Based

Capital Guidelines: Market Risk, Alternatives to Credit Ratings for Debt and Securitization Positions" under the supervision of the Department of the Treasury Office of the Comptroller of the Currency, Federal Reserve system and the Federal Deposit Insurance Corporation (known collectively, as "the agencies") and that have proposed to incorporate into the Proposed Market Risk Capital Rules certain alternative methodologies for calculating specific risk capital requirements for debt and securitization positions that do not rely on credit.[10]

The Board of Governors of the Federal Reserve System's Comprehensive Capital Analysis Review (CCAR) involved the Fed's evaluation of the internal capital planning processes of large, complex bank holding companies and their proposals to undertake capital action in 2011. The Federal Reserve issued guidelines on November 17, 2010, to provide a common approach to ensure that the bank holding companies hold adequate capital to maintain ready access to funding, continue operations and meet their obligations to creditors and counterparties, and continue to serve as credit intermediaries, even under adverse conditions. The following nineteen large bank holding companies submitted comprehensive capital plans and additional supervisory information to the Federal Reserve in January 2011.[11]

The preceding list of nineteen bank holding companies that participated in the 2011 CCAR are bound by the Dodd-Frank Wall Street Reform and Consumer Protection Act, Title I, Sec. 165, that states in part that the Board of Governors shall, on its own or pursuant to recommendations by the Council under Section 115, establish prudential standards for nonbank financial companies supervised by the Board of Governors and bank holding companies with total consolidated assets equal to or greater than $50,000,000,000—. A key innovation in the CAR is the expectation that large bank holding com-

Table 6.6. The 19 Bank Holding Companies in the 2011 Comprehensive Capital Analysis and Review (CCAR) [1]

Ally Financial, Inc.	American Express Company
Bank of American Corporation	The Bank of New York Mellon
BB&T Corporation	Citigroup, Inc.
Capital One Financial Corporation	The Goldman Sachs Group, Inc.
Fifth Third Bancorp.	Keycorp
JPMorgan Chase & Co.	Morgan Stanley
MetLife, Inc.	Regions Financial Corporation
The PNC Financial Services Group	SunTrust Bank, Inc.
State Street Corporation	Wells Fargo & Company
U. S. Bancorp	

[1] Ibid., note

Tabale 6.7. The 19 Bank Holding Companies in the 2012 Comprehensive Capital Analysis and Review (CCAR)

Ally Financial Inc.	American Express Company
Bank of America Corporation	The Bank of New York
BB&T Corporation	Citigroup, Inc..
Capital One Financial Corporation	The Goldman Sachs Group, Inc.
Fifth Third Bancorp	Keycorp
JPMorgan Chase & Co. Inc.	Morgan Stanley
MetLife, Inc.	Regions Financial
The PNC Financial Services	SunTrust Bank, Inc.
State Street Corporation	Wells Fargo & Company
U. S. Bancorp	

panies submit annual comprehensive capital plans to the Federal Reserve. These plans will describe their strategies for management of their capital over a twenty-four-month, forward planning horizon.[12]

The Board of Governors of the Federal Reserve, in its Comprehensive Capital Analysis and Review 2012: Methodology for Stress Scenario Projections, March 12, 2012, emphasized that the CCAR is a supervisory assessment by the Federal Reserve of the capital planning processes and capital adequacy of these large, complex bank holding companies (BHCs). The CCAR is the Federal Reserve's central mechanism for developing supervisory assessments of capital Adequacy at these firms. The nineteen immediately preceding bank holding companies were required to participate in this year's CCAR (CCAR 2012) and they submitted comprehensive capital plans to the Federal Reserve, describing their strategies for managing their capital over a nine-quarter planning horizon.

The Federal Reserve Board on November 15, 2012, released the economic and financial market scenarios that will be used in the next round of stress

Table 6.8. The 19 Bank Holding Companies in the 2013 Comprehensive Capital Analysis and Review (CCAR)

Ally Financial Inc.	American Express Company
Bank of America Corporation	The Bank of New York Mellon Corporation
BB&T Corporation	Citigroup, Inc.
Capital One Financial Corporation	The Goldman Sachs Group, Inc.
Fifth Third Bancorp	Keycorp
JPMorgan Chase & Co.	Morgan Stanley
MetLife, Inc.	Regions Financial Corporation
The PNC Financial Services Group, Inc.	SunTrust Banks, Inc.
State Street Corporation	Wells Fargo & Company
U. S. Bancorp	

tests for large financial institutions. The scenarios include baseline, adverse, and severely adverse scenarios, as described in the Federal Reserve's final rules that implement stress test requirements of the Dodd-Frank Wall Street Reform and Consumer Protection Act. Each scenario includes 26 variables, including economic activity, unemployment, exchange rates, prices, incomes, and interest rates.

For the nineteen bank holding companies that are part of the Federal Reserve's CCAR, the three scenarios will be used in stress tests conducted under Dodd-Frank Act stress test rules and for the CCAR, including company-run stress tests and supervisory stress tests conducted by the Federal Reserve, (http://www.federalreserve.gov/newsevents/press/bcreg/20121115a.htm) In April 2012, the J. P. Morgan & Company Chief Investment Office (CIO) London had the responsibility for the management of $350 billion in excess deposits, in fact, placed a huge bet on a complex set of synthetic credit derivatives that lost at least $6.2 billion in 2012. At the Chief Investment Office (CIO), London, huge losses occurred as a result of the so-called "London Whale" volume of trading that was executed by traders in the magnification and huge size of the trades.

The trading losses doubled rapidly, according to the U.S. Senate Permanent Subcommittee on Investigations, Committee on Homeland Security and Governmental Affairs, JPMorgan Chase Whale Trades: A Case History of Derivatives Risks and Abuses,[13] March 15, 2013, and then tripled, (http://www.hsgac.senate.gov/subcommittees/investigations). The huge $6.2 billion in losses also rendered obsolete the Federal Reserve Comprehensive Capital Analysis and Review (CCAR); shocked the investing public; placed Emphasis on the Chief Investment Office, London, that was responsible for bankrolling high stakes, high risk credit derivatives trade; authenticated the manner in which the JPMorgan Chase "Whale Trades" provide an startling and instructive case history in which the synthetic credit derivatives have become a multibillion dollar source of risk within the U.S. banking system; exposed broader, systemic problems related to the valuation, risk analysis, disclosure and oversight of synthetic credit derivatives that are in the possession of U.S. financial institutions.

The U.S. Senate Permanent Subcommittee on Investigations Staff Report, JPMorgan Chase Whale Trades: A Case History of Derivatives Risks and Abuses, March 15, 2013, affirmed that the JPMorgan Chase Whale Trades history showed how JPMorgan Chase outmaneuvered its regulators by keeping the high risk Synthetic Credit Portfolio off the radar of the Comptroller of the Currency despite the massive size of the $6.2 billion losses and the three months of escalating losses; routing bank reports that might have drawn attention to the Synthetic Credit Portfolio (SCP) were delayed, detailed data were omitted, blanket assurances were offered when they should not have been, and requested information was late or not provided at all.[14]

The volume of trading activity that is the center of this JPMorgan Chase Whale Trades focuses on complex credit derivatives that include credit default swaps, credit indices, and credit index tranches. Derivatives, by definition, refer to the financial contract or instrument whose price is determined or derived from the value of an underlying asset, rate, entity, weather, index, ratio, or event. In essence, the credit derivatives place bets whether, during a specified period of time, the referenced financial contracts, instruments or entities will experience a negative credit event, such as a bankruptcy, default, or failure to pay. The parties taking the "long" side of the bet, then wager that no credit event will occur; the parties taking the "short" side of the bet, then wager that the negative credit event will occur. These credit instruments are often described as "synthetic" because they do not contain any tangible assets such as a loan or bond; they simply reference the financial instrument or entity whose credit quality is at issue.[15]

The credit default swap within the JPMorgan Chase Whale Trades: A Case History of Derivatives Risks and Abuses, is the simplest type of credit derivative that also dominates the credit derivative markets; a credit default swap (CDS) is a contract between two parties who are placing opposite bets on the 247 creditworthiness of a specified financial instrument, entity or event. Traders often connect credit default swaps to insurance contracts; the "long" party is essentially selling insurance, or credit protection, against the occurrence of a negative credit event, while the "short" party is essentially buying that insurance or credit protection. Generally, the credit default swaps are entered into between a swap dealer and an institutional investor like a hedge fund, insurance company or other financial institution.

There is an increasingly complicated form of credit derivative that includes a credit index. Most significantly, the credit indices were first invented by JPMorgan Chase and Morgan Stanley in 2001 and each credit index references a basket of selected credit instruments such as the credit default swaps or other types of credit instruments. The credit index volume of transactions are typically entered into over-the-counter (defined as outside of a regulated exchange) between a swap dealer and an investor.[16]

The JPMorgan Chase Whale Trades Chief Investment Office (CIO) traded a variety of credit indices (IG9, HY, and iTraxx Indices) and held more than one hundred different types of credit derivative instruments by March 2012. The CIO's largest holdings involved indices under the management of the Markit Group, Ltd. and the latter is a global financial information services company that is in the management of multiple index products; Markit Group, Ltd., in fact, owns and operates the indices while the firm performs a variety of services that are related to multiple index products, among which are calculating the index values and publishing the daily index prices on its website.[17]

More particularly, the IG9 5-Year has the meaning that the swap referencing the IG9 index will expire in 2012, five years after the IG9 index was issued. The HY refers to High Yield because the index tracks the credit default swaps that name 100 North American companies that pose higher credit risks and so produce higher returns to investors; these companies are often rated as "HY" companies because they carry non-investment grade or "Junk Bond" ratings. The JPMorgan Chase Whale Trades Chief Investment Office (CIO) transactions involved trading in the iTraxxEurope index which tracks credit default swaps for 125 investment grade companies in Europe. A still more complicated type of credit derivative involved credit tranches. The CIO traded credit tranches typically related to the Markit Group, Ltd. Credit indices. Each of the Markit Group, Ltd., credit indices tracked the value of a specified basket of credit instruments. Instead of requiring bets on the creditworthiness of the entire basket, for some credit indices, Markit Group offered instruments that enabled parties to place bets on just a portion of the basket, offering four tranches with different degrees of vulnerability to default. The richest tranche, called the "equity tranche," was immediately affected by any default at any company in the basket. The next tranche, called the "mezzanine," was affected only by losses that exceeded 15 percent of the loss distribution. Those losses usually required one or more defaults to take place. The next tranche, called the "senior tranche," was affected only by losses that exceeded 25 percent of the loss distribution. The last and most secure tranche, called the "super senior tranche," was affected only by losses that exceeded 35 percent of the loss distribution. Those losses typically required multiple defaults to take place.[18]

There are the "Findings of Fact" detailed in the JPMorgan Chase Whale Trades: A Case History of Derivatives Risks and Abuses Majority and Minority Staff Report by The Permanent Subcommittee on Investigations, U.S. Senate The largest U. S. banks in 2013 are reported deep into derivatives, complex financial instruments that derive their value from other assets, events, entities, and indices. The derivatives behind the JPMorgan Chase Whale Trades were part of a so-called Synthetic Credit Portfolio that essentially made outsized bets on whether particular financial instruments, events or entities were creditworthy or would default during specified time periods. The Permanent Subcommittee on Investigations Staff Report—JPMorgan Chase Whale Trades includes the following among its findings of fact,[19] from the Permanent Subcommittee on Investigations, (a) in the first quarter of 2012, without alerting its regulators, JPMorgan Chase's Chief Investment

Office used bank deposits, including some that were federally insured, to construct a $157 billion portfolio of synthetic credit derivatives, engaged in high risk, complex, short-term trading strategies, and disclosed the extent and high risk nature of the portfolio to its regulators only after it attracted media attention, (b) JPMorgan Chase, through its Chief Investment Office, hid over $660 million in losses in the Synthetic Credit Portfolio (SCP) for several months in 2012, by allowing the CIO to overstate the value of its credit derivatives; ignoring red flags that the values were inaccurate, including conflicting Investment Bank values and counterparty collatr4al disputes and supporting reviews which exposed the Synthetic Credit Portfolio's questionable pricing practices but upheld the suspect values, © JPMorgan Chase dodged the Office of the Comptroller of the Currency (OCC) oversight of the Synthetic Credit Portfolio (SCP) by not alerting the OCC to the nature and extent of the portfolio; failing to inform the OCC when the SCP grew tenfold in 2011 and tripled in 2012; omitting SCP specific data from routine reports sent to the OCC; omitting mention of the SCP's growing size, complexity, risk profile, and losses, and (d) in the first three months of 2012, JPMorgan Chase disregarded the warning signals and downplayed the SCP's risk by allowing the CIO to raise the limits, change its risk evaluation models and continue trading.

The "Findings of Fact" in the JPMorgan Chase Whale Trades provide documentation that strengthened the unfavorable restraint of trade practices that form the foundation of the financial and economic crisis and economic collapse 2007–2008 and the JPMorgan Chase Whale Trades 2012–2013. Senator Carl Levin, Chairman, U.S. Senate Permanent Subcommittee on Investigations, Homeland Security and Governmental Affairs Committee, in his opening statement of the Hearing on JPMorgan Chase Whale Trades: A Case History of Derivatives Risks and Abuses, March 15, 2013, delivered a strong statement that JPMorgan Chase is the largest financial holding company in the United States; JPMorgan Chase & Company is also the largest derivatives dealer in the world, in addition to being the largest single participant in the world credit derivatives markets.

Most significantly, the staff report, JPMorgan Chase Whale Trades: A Case History of derivatives Risks and Abuses, established without a doubt that JPMorgan Chase & Company (JPMorgan Chase) is a leading global financial services firm incorporated under Delaware law and headquartered in New York City. In addition to being the largest financial holding company in the United States, it conducts operation in more than 60 countries, employs more than 240,000 people, maintains 5,500 bank branches, and as of December 31, 2012, has more than $2 trillion in assets.[20]

In essence, JPMorgan Chase is also the largest derivatives dealer in the United States, active in derivatives markets involving commodities, credit instruments, equities, foreign currencies, and interest rates. Four U.S. banks dominate the U.S. derivatives markets, of which the credit derivatives market is the third largest, representing about 6 percent of all derivatives activities. JPMorgan Chase is the largest U.S. derivatives dealer in the credit markets.

NOTES

1. Economic Report of the President Transmitted to the Congress February 2012 Together With the annual Report of the Council of Economic Advisers, "Money Stock, Credit, and Finance, Table B-69. Money Stock and Debt Measures, 1972–2011, p. 399," Washington, DC: Superintendent of Documents, U.S. Government Printing Office, 2012, p. 399

2. Letter from Trabue Bland, Vice President, Regulatory Affairs and Assistant General Counsel, Intercontinental Exchange, Inc., 2100 RiveEdge Parkway Ste. 500, Atlanta, GA, 30328, To: Federal Deposit Insurance Corporation, Securities and Exchange Commission, Board of Governors of the Federal Reserve System, Department of the Treasury Office of the Comptroller of the Currency, RE.: Restrictions on Proprietary Trading and Certain Relationships With Hedge Funds ("Volcker"), dated February 12, 2012.

3. This table is a compilation from the Board of Governors of the Federal Reserve System, Flow of Funds Accounts of the United States—FFA Coded Tables, December 6, 2013, FFA Coded Tables, June 8, 2006, B.100 Balance Sheet of Households and Nonprofit Organizations (Sector includes domestic Hedge Funds, Private Equity Funds, and Personal Trust), Washington, DC.

4. Federal Reserve Bulletin, "Changes in U.S. Family Finances from 2007 to 2010: Evidence from the Survey of Consumer Finances," by Jesse Bricker, Arthur B. Kennickell, Kevin B. Moore, and John Sabelhaus of the Board's Division of Research and Statistics, June 2012, Vol. 98, No. 2, pp. 1–80.

5. Ibid., p. 1.

6. Ibid., p. 1.

7. Ibid., pp. 1–2.

8. Wall Street ad the Financial Crisis: Anatomy of a Financial Collapse, Report and Appendix, Before the Permanent Subcommittee on Investigations of the Committee on Homeland Security and Governmental Affairs, U.S. Senate, 112th Congress, First Session, Volume 5 of 5-PART-I, April 13, 2011, p. 59.

9. Ibid., pp. 64 and 65.

10. Federal Register, Vol. 76, No. 245, Part IV, Department of the Treasury Office of the Comptroller of the Currency, Federal Reserve System; Federal Deposit Insurance Corporation, 12 CFR Parts 3, 208, 225et al, Risk-Based Capital Guidelines: Market Risk; Alternatives to Credit Ratings for Debt and Securitization Positions; Proposed Rule, December 21, 2011, p. 79380; Department of the Treasury Office of

the Comptroller of the Currency, 12 CFR Part 3, Docket ID: OCC—2012-0002, RIN 1557-AC99; Federal Reserve System 12 CFR Parts 208 and 225, Regulations H and Y; Docket No. R-1401; RIN No. 7100-AD 61; Federal Deposit Insurance Corporation, 12 CFR Part 325 RIN 3064-AD70, Risk-Based Capital Guidelines: Market Risk; Action: Joint Final Rule; Dates: the Final Rile is effective January 1, 2013.

11. Board of Governors of the Federal Reserve System, Comprehensive Capital Analysis Review: Objectives and Overview, March 18, 2011, Executive Summary, p. 1.

12. Ibid., pp. 3 and 4.

13. U.S. Senate Permanent Subcommittee on Investigations of the Committee on Homeland Security and Governmental affairs, JPMorgan Chase Whale Trades: A Case History of Derivatives Risks and Abuses, Majority and Minority Staff Report, Permanent Subcommittee n Investigations, United States Senate Released in Conjunction With the Permanent Subcommittee on Investigations, March 15, 2013 Hearing, pp. 2–14.

14. Ibid., p. 14.

15. Ibid., p. 29.

16. Ibid., pp. 30–31.

17. Ibid., p. 31.

18. Ibid., p. 33.

19. Ibid., pp. 15 and 16.

20. Ibid., pp. 18 and 19.

Bibliography

PUBLIC DOCUMENTS

95th Congress, 2d Session, U.S. Senate, *Interlocking Directorates Among the Major U.S. Corporations*, Staff Study prepared by the Subcommittee on Reports, Accounting, and Management of the Committee on Governmental Affairs, Document No. 95-107, June 15, 1978, Washington, DC: U.S. Government Printing Office, 1978.

96th Congress, 2d Session, *Structure of Corporate Concentration: Institutional Shareholders and Interlocking Directorates Among Major U.S. Corporations*, Staff Study, Committee on Governmental Affairs, U.S. Senate, vol. 1, December 1980, Washington, DC: U.S. Government Printing Office, 1980.

106th Congress, 1st Session, House of Representatives, Gramm-Leach-Bliley 109th Congress 2d Session, U.S. Senate, *The Role of Market Speculation in Rising Oil and Gas Prices: A Need to Put the Cop Back on the Beat*, Staff Report prepared by the Permanent Subcommittee on Investigations of the Committee on Homeland Security and Governmental Affairs, June 27, 2006, Washington, DC: U.S. Government Printing Office, 2006.

111th Congress, 2d Session, House of Representatives, Conference Report, *Dodd-Frank Wall Street Reform and Consumer Protection Act*, Report 111-517, June 29, 2010.

Act, Report 106-434, November 2, 1999.

Bank Holding Company Act of 1956 (12 U.S.C. 1843).

Board of Governors of the Federal Reserve System, "Financial Holding Companies," Washington, DC, May 18, 2012.

Board of Governors of the Federal Reserve System, "Industrial Production and Capacity Utilization (G.17)," 2012.

Board of Governors of the Federal Reserve System, *The Federal Reserve System: Purposes & Functions*, 9th ed., Washington, DC, 2005.

Bureau of Economic Analysis (BEA), *Concepts and Methods of the U.S.* Chairman Ben S. Bernanke, Board of Governors of the Federal Reserve System, Testimony: "Causes of the Recent Financial and Economic Crisis," before the Financial Crisis Inquiry Commission (FCIC), September 2, 2010, Washington, DC.

Chairman Gary Gensler, Commodity Futures Trading Commission, "Keynote Address of Chairman Gary Gensler, OTC Derivatives Reform, Market's Outlook for OTC Derivatives Markets Conference," Washington, DC, 2010.

Commercial Banks and Their Trust Activities: Emerging Influence on the American Economy, vol. 1, Staff Report for the Subcommittee on Domestic Finance, Committee on Banking and Currency, House of Representatives, 90th Congress, 2d Session, July 8, 1968, Washington, DC: Government Printing Office, 1968.

Economic Report of the President, Transmitted to the Congress February 2012 Together With the Annual Report of the Council of Economic Advisers, "Money Stock, Credit, and Finance, Table B-69," Washington, DC: U.S. Government Printing Office, 2012.

Executive Office of the President, Office of Management and Budget, *North American Industry Classification System: United States, 2012*, Lanham, MD: Bernan Press, 2012.

Federal Reserve Bank, "Large Commercial Banks: Insured U.S.-Chartered Commercial Banks that Have Consolidated Assets of $300 Million or More, Ranked by Consolidated Assets as of March 31, 2012," Federal Reserve Statistical Release, March 31, 2012.

Federal Reserve Bulletin, "Changes in U.S. Family Finances from 2007 to 2010: Evidence from the Survey of Consumer Finances," Jesse Bricker, Arthur B. Kennickell, Kevin B. Moore, and John Sabelhaus of the Board's Division of Research and Statistics, June 2012, vol., 98, no. 2.

Federal Reserve Statistical Release, H.6. (508), "Money Stock Measures," August 23, 2012, Table 1: Money Stock Measures.

Financial Crisis Inquiry Commission, *The Financial Crisis Inquiry Report Final Report of the National Commission on the Causes of the Financial and Economic Crisis in the United States*, Official Government Edition, submitted by the Financial Crisis Inquiry Commission Pursuant to Public Law 111-21, January 2011.

Funds, Leverage, and the Lessons of Long-Term Capital Management, Department of the Treasury, Board of Governors of the Federal Reserve System, Securities and Exchange Commission, Commodity Futures Trading Commission, Washington, DC, April 1999.

GAO Report to Congressional Requesters, *Long-Term Capital Management: Regulators Need to Focus Greater Attention on Systemic Risk*, GAO/GGD-00-3, Washington, DC, October 1999.

Library of Congress Congressional Research Service, "The Exercise of Voting Rights by Large Institutional Investors: A Survey," prepared by Julius W. Allen, Consultant, Economics Division, December 2, 1977.

Money Trust Investigation: Investigation of Financial and Monetary Conditions in the United States Under House Resolutions Nos. 429 and 504 Before a Subcom-

mittee on Banking and Currency: Interlocking Directorates, *National Income and Product Accounts*, Washington, DC, November 2010.

National Nanotechnology Initiative: Leading to the Next Industrial Revolution: A Report by the Interagency Working Group on Nanoscience, Engineering and Technology (IWGN), Committee on Technology, National Science and Technology Council, Washington, DC, February 2000.

National Science and Technology Council Committee on Technology, Subcommittee on Nanoscale Science, Engineering, and Technology, *National Nanotechnology Initiative Strategic Plan*, Washington, DC: Executive Office of the President, February 2011.

Over-the-Counter [off the exchanges] Derivatives Markets and the Commodity Exchange Act: Report of the President's Working Group on Financial Markets, Department of the Treasury, Board of Governors of the Federal Reserve System, Securities and Exchange Commission, Commodity Futures Trading Commission, November 1999.

Report of the President's Working Group on Financial Markets, on Hedge

Summers, Larry H., Secretary, U.S. Department of the Treasury; Alan Greenspan, Chairman, Board of Governors of the Federal Reserve System; Arthur Levitt, Chairman, U.S. Securities and Exchange Commission; William J. Rainer, Chairman, U.S. Commodity Futures Trading Commission, *"Over-the-Counter Derivatives Markets and the Commodity Exchange Act*, Report of the President's Working Group on Financial Markets, November 1999.

U S. Senate Permanent Subcommittee on Investigations Hearing on JPMorgan Chase Whale Trades: A Case History of Derivatives Risks and Abuse, March 15, 2013.

U.S. Senate Permanent Subcommittee on Investigations, Homeland Security and Governmental Affairs Committee, Opening Statement of Senator Carl Levin (D-Mich.), U.S. Census Bureau, "Statistical Abstract of the United States: 2012," 131st Edition, Washington, DC, 2011.

U.S. Congressional House Committee on Banking and Currency, *Report of the Committee Appointed Pursuant to House Resolutions 429 And 504 to Investigate the Concentration of Control of Money and Credit*, February 28, 1913 (Reprint of the 1913 ed.), Arno Press, New York, 1975.

U.S. General Accounting Office (GAO) Report to Congressional Committees, *Financial Derivatives Actions Taken or Proposed Since May 1994* (GAO/GGD/AIMD-97-8), Washington, DC, November 1996.

U.S. Senate, 76th Congress, 3d Session, *Investigation of Concentration of Economic Power—Temporary National Economic Committee*, Monograph no. 29, "The Distribution of Ownership in the 200 Largest Nonfinancial Corporations" (Printed for the use of the Temporary National Economic Committee), Washington, DC: U.S. Government Printing Office, 1940.

U.S. Senate, 95th Congress, 2d Session, *Voting Rights in Major Corporations*, Staff Study prepared by the Subcommittee on Reports, Accounting, and Management of the Committee on Governmental Affairs, January 1978, Document No. 95-99, Washington, DC: U.S. Government Printing Office, 1978.

United Nations International Standard Industrial Classification of All Economic Activities, Rev. 4.

U.S. General Accounting Office Report to Congressional Committees, *Financial Derivatives Actions Taken or Proposed Since May 1994*, GAO/GGD/AIMD-97-8.

U.S. Senate, Permanent Subcommittee on Investigations, Committee on Homeland Security and Governmental Affairs, *JPMorgan Chase Whale Trades: Risks And Abuses*, Majority and Minority Staff Report, Permanent Subcommittee on Investigations, March 14, 2013.

Wall Street and the Financial Crisis: Anatomy of a Financial Collapse, Report and Appendix Before the Permanent Subcommittee on Investigations of the Committee on Homeland Security and Governmental Affairs, U.S. Senate, 112th Congress, 1st Session, vol. 5, Part II, April 13, 2011.

Washington, DC: Government Printing Office, 1918.

ASSOCIATIONS AND BANKS

Allied Social Sciences Association
American Economic Association
Association of Comparative Economic Studies
Association of Evolutionary Economics
Association of Social Economics
Bank for International Settlements
Central Bank of Brazil
Central Bank of Japan
Central Bank of Russia
Central Bank of South Africa
Economic History Association
Industrial Organization Society
National Economic Association
The People's Bank of China
World Economics Association

OTHER DOCUMENTS

Bonbright, James C., and Gardner C. Means, *The Holding Company: Its Public Significance and Its Regulation*, 1st ed. (reprint ed., 1969, Augustus M. Kelley Publishers, New York, NY), New York: McGraw-Hill Book Company, Inc., 1932.

Curcuru, Stephanie E., "U.S. Cross Border Derivatives Data: A User's Guide," Federal Reserve Bulletin Board, Division of International Finance, May 2007.

Dunbar, Nicholas, *The Devil's Derivatives: The Untold Story of the Slick Traders and Hapless Regulators Who Almost Blew Up Wall Street . . . and Are Ready to Do It Again*, Boston, MA: Harvard Business Review Press, 2011.

Harris, Dr. James A., Sr., *Nanotechnology World Economy*, Bloomington, IN: AuthorHouse, 2009.

InterContinentalExchange, Inc., Letter from Trabue Bland, Vice President, Regulatory Affairs and Assistant General Counsel, InterContinentalExchange, Inc., 2100 RiverEdge Parkway Ste. 500, Atlanta, GA 30328, to Federal Deposit Insurance Corporation; Securities and Exchange Commission; Board of Governors of the Federal Reserve System; Department of the Treasury, Office of the Comptroller of the Currency, re "Restrictions on Proprietary Trading and Certain Relationships with Hedge Funds" ("Volcker"), February 12, 2012.

Ivry, Bob, Bradley Keoun, and Phil Kuntz, "Secret Fed Loans Helped Banks Net $13 Billion," *Bloomberg Markets*, November 27, 2011.

Kosar, Kevin R., *CRS Report for Congress: Government-Sponsored Enterprise (GSEs): An Institutional Overview*, prepared for Members and Committees of Congress, Order Code RS 21663, updated April 23, 2007.

National Institute for Materials Science (NIMS), "Making Full Use of the 'Eyes' and 'Hands' of Nanotechnology," *Nanotechnology Platform Japan*, November 2012.

National Institute for Materials Science (NIMS), "Nano-Bio: Medical Science Meets Nanoarchitectonics," *NIMS NOW International*, no. 4, May 2012.

National Institute for Materials Science (NIMS), "Seeing the Nanoworld NIMS Transmission Electron Microscopy," *NIMS NOW International*, no. 5., June 2013.

National Institute for Materials Science (NIMS), "Impact of NIMS Research Results on Society," vol. 11, no. 10, December 2013.

Reconstruction Finance Corporation Act as Amended and Provisions of the Emergency Relief and Construction Act of 1932 Pertaining to the Reconstruction Finance Corporation, July 21, 1932, Washington, DC: U.S. Government Printing Office, 1932.

JOURNALS AND PERIODICALS

American Economic Review

Development and Socio-Economic Progress

Eastern Economic Journal

Economic Challenger: A Quarterly Journal of Economics, Commerce and Management

International Social Science Journal (UNESCO)

Journal of International Business Studies

Journal of the American Political Science Association

Journal of the World Economics Association

Real-World Economics Review

Review of International Industrial Organization

World Economic Review

Index

About the Author

James A. Harris Sr., PhD, is an economist, professor, business executive, author, registered individual international consultant with the United Nations Global Marketplace (UNGM), United Nations Economic Commission for Africa (UNECA) Database of African Experts, and the Diaspora. He originated a development proposal on the construction of an oil refinery and manufacture of petroleum products while appointed Commissioner, Economic Development Authority of the City of Portsmouth, Virginia, and the Portsmouth Port and Industrial Commission.

Dr. Harris was adjunct professor of management, New York University School of Continuing and Professional Studies, 1982 to 2005, where he was elected as recipient of the Great Teacher Award by the Alumni Federation of New York University and Medallion Society in 1994, one of New York University's most prestigious honors in recognition of superb accomplishment; for outstanding teaching, New York University School of Continuing and Professional Studies Faculty Longer Service Recognition 20 Years Service Award in April 2002; New York University School of Continuing Education Award for Teaching Excellence and Outstanding Talent as a Teacher in 1986; and New York University School of Continuing Education, Division of Degree Studies, 1985–1986 Award for Outstanding Teaching.

For fifteen years up to 1983, he was an economist/executive of the American Telephone and Telegraph Company (AT&T) Headquarters State Regulatory Matters Department and the Comptrollers Department–Business Research Division. He was appointed Consultant and Team Leader by the United Nations Development Programme (UNDP) Bureau of Africa for November 1989 to December 1989. He was appointed Economist, First National City Bank Economics Department, from 1969 and 1970 and Business

Economist, U.S. Federal Trade Commission, Bureau of Economics; he conducted industrial organization economics studies and reports that evaluated economic evidence in antitrust assessment and restraint of trade.

Dr. Harris received his AB from Virginia State University, Petersburg, VA; his MA from New York University Graduate School of Arts and Science, New York; and his PhD from Pacific Western University, Los Angeles, in Economics with honors in 1983.

He is the author of *Nanotechnology World Economy*, Bloomington, Indiana: AuthorHouse, 2009, and the author of other books and numerous articles.

CPSIA information can be obtained at www.ICGtesting.com
Printed in the USA
BVOW04*1333161014

370656BV00010B/7/P

9 780739 199985